Solid State Devices 1982

ESSDERC-SSSDT Meeting at Munich,
13-16 September 1982

Edited by A. Goetzberger and M. Zerbst

physik-
verlag

Solid State Devices 1982

ESSDERC-SSSDT Meeting at Munich,
13-16 September 1982

Edited by A. Goetzberger and M. Zerbst

 physik-verlag Weinheim

PHYSICS

Prof. Dr. A. Goetzberger
Fraunhofer-Institut
für Solare Energiesysteme
Oltmannsstr. 22
D-7800 Freiburg

Dr. M. Zerbst
Siemens AG
Otto-Hahn-Ring 6
D-8000 München 83

Publisher's editor: Hans F. Ebel
Production manager: Peter J. Biel

This book contains 129 figures and 9 tables

Deutsche Bibliothek Cataloguing-in-Publication Data
CIP-Kurztitelaufnahme der Deutschen Bibliothek
[Solid state devices nineteen hundred and eighty-two]
Solid state devices 1982 / ESSDERC SSSDT meeting at Munich, 13 – 16. September 1982.
Ed. by A. Goetzberger and M. Zerbst. [Sponsored by the Europ. Physic. Soc. (EPS) ...].
– Weinheim: Physik-Verlag, 1983.
  ISBN 3-87664-072-5
NE: Goetzberger, Adolf [Hrsg.]; ESSDERC ⟨12, 1982, München⟩; SSSDT ⟨07, 1982, München⟩

Printing: Krebs-Gehlen Druckerei GmbH & Co. KG, D-6944 Hemsbach
Bookbinding: Wilh. Osswald & Co., D-6730 Neustadt
Printed in the Federal Republic of Germany

Preface

The field of solid state devices continues to show a broad spectrum of activities. The 12th European Solid State Device Research Conference (ESSDERC) and the 7th Symposium on Solid State Device Technology (SSSDT) were held jointly on 13 – 16 September 1982 in Munich. Together they featured invited lectures, of which eight are published in this volume. In addition, there were 119 contributed papers.

The papers published here mirror some of the most active areas of development. Photovoltaic energy conversion, which is surveyed in the first paper, has grown to such proportions that specialized meetings on this topic are being arranged. Nevertheless, the organizers of ESSDERC considered it worthwhile to present a review about the latest results. In the same way, the technology for high speed GaAs VLSI, metal-organic vapor phase epitaxy, thin oxide gate technology and advances in silicon materials preparation are dealt with. One of the major concerns today is microstructure lithography. The paper on synchrotron lithography is a timely account. Also sensors continue to receive much attention, as can be deduced from the contribution on the sensor lag. And finally there is a paper on surface acoustic wave devices and infrared detectors with integrated signal processing.

The editors would like to thank all contributors for their collaboration. They hope that this volume will be of use to many readers.

Freiburg and Munich, December 1982

A. Goetzberger
M. Zerbst

Sponsored by:
European Physical Society (EPS)
Institute of Physics (IOP)
Region 8 of the Institute of Electrical and
Electronics Engineers (IEEE)
Deutsche Physikalische Gesellschaft (DPG)
Nachrichtentechnische Gesellschaft im VDE (NTG)

Contents

The Status of Photovoltaic Devices

Sigurd Wagner

Department of Electrical Engineering and Computer Science, Princeton University, Princeton, New Jersey 08544, USA.

SUMMARY

Today's commercial photovoltaic devices are planar n^+pp^+ silicon diodes made of single crystals, cast polycrystalline ingots, or ribbons. Power conversion efficiencies, η, range from 8 to 16 percent. Costs are being reduced and efficiencies raised by introduction of alternative silicon materials, refined device structures and new processing steps. Yet, conventional silicon cells may not attain the efficiency-to-cost ratio, γ, required for utility applications. Therefore, a bewildering diversity of device structures and cell materials is under research. One distinguishes between concentrator cells which increase γ by raising η, and thin-film cells which increase γ by reducing the cost.

The development of silicon concentrator cells includes device design (horizontal or vertical junctions, interdigitated contacts) and advanced processing accompanied by refinement of device theory. Eventually such Si cells may reach $\eta \simeq 25$ % in concentrated sunlight. For concentrator cells made of compound semiconductors the focus has shifted from single diodes to monolithic cascaded cells where aliquot portions of the photon flux are converted in semiconductors with different band gaps. Such cascade cells rely on con-

ventional heteroface junctions. Present emphasis lies on the preparation of lattice-matching alloys with proper energy gaps, and of
connecting tunnel junctions. Cascade cells may reach $\eta \geq 30$ %.

Thin-film cells are least developed; research is heavily materials-
oriented. However, the attention given to hydrogenated amorphous
silicon has brought this particular semiconductor to a stage where
device and materials research are equally important for efficiency
improvement. Single thin film cells may be expected to reach
$\eta \cong 15$ %. Laboratory efficiencies for several competing materials
lie in the vicinity of 10 %.

Using selected devices, I shall discuss the status of photovoltaic
device technology as well as the current directions of research.

1. INTRODUCTION

Production of mass electricity is the most ambitious commercial goal
for the development of photovoltaic systems /1/. Photovoltaic converters must become competitive with conventional technologies for
the generation of electric power to reach a market of utility scale.
The competitive, break-even, price for photovoltaic electricity may
vary with country, location, owner as well as economic an political
conditions. However, there is no question that the goal of providing
electricity in a utility-like mode calls for an increase in cell performance and for drastic systems cost reductions beyond those achieved to date.

Special markets, aside from that for large scale power generation,
do exist for photovoltaics. Space vehicles and users remote from
electric grids are examples. Space applications demand particularly high conversion efficiency, high power to weight ratio and radiation hardness; cost is a secondary factor. Remote generation of power
may tolerate moderate efficiency at substantial cost. Yet, these
special markets are small. Power generation on a large scale is the
goal of the extensive photovoltaic research and development programs
now carried out in the United States, in Western Europe and in Japan.

A kilowatt hour price of 15 ¢/kWh translate to a cost of $2,000/kW$_{peak}$ (a.c.) for the installed system. To reach this cost the system efficiency will have to be 12 %, with a flat plate module efficiency of 15 % (at T =25 C) and a module cost of about $700/kW$_{peak}$ (d.c. at T = 25 C). The cell efficiency will have to be higher than 15 %. Today modules, based on single crystal silicon, with a conversion efficiency of 9 to 10 % can be purchased in quantity at $6,000/kW$_p$.

It is obvious that solar cell research must address both raising the efficiency and reducing the cost. Extensive discussions have been held about the relative importance of these two criteria. Suffice it to point out that cost reduction must be shared proportionally between the various components of a photovoltaic system. On the other hand, the burden of efficiency increase will have to be carried just by the module.

Are concentrators more promising than flat plates? Concentrator modules will need cells with an efficiency of 20 % or higher at a cost of ~$1/cm^2 . Some experts favor concentrators, arguing that this cost target already has been reached with today's integrated circuits, and that the improvement of performance is well within the realm and experience of semiconductor technology.

In our technical discussion we shall attempt to provide a balance between the more promising devices. We will set out with a description of the present silicon cell technology for flat plate modules. This description can be extended conveniently to the various concentrator cell designs based on silicon. The practical efficiency limit of silicon cells in concentrated sunlight is thought to lie below but close to 25 %. Spectral splitting techniques employing two or more cells with different band gaps made of different semiconductor materials may reach 30 % or more under concentrations. Such cells we will dicuss next.

Thin film cells are expected to be highly amenable to cost reduction. In fact, their potential for low cost is taken for granted while much attention has been paid during the past few years to reaching laboratory efficiencies of 10 %. Four materials have reached this milestone. A bewildering variety of semiconductors is under investigation, but a convenient classification is that in polycrystalline and in amorphous cells. This is the order in which we shall organize our survey.

Ample literature is available for the reader who seeks an introduction to photovoltaics /3-6/. The proceedings of the sesquiannual IEEE conference on photovoltaics are a rich source of information for the specialist /7/.

2. THE SILICON CELL

The silicon cell is an n^+p diode with a shallow (0.1 to 1 μm) phosphorus diffusion in a 1 to 10 Ω cm p-type boron-doped 300 μm thick substrate. The p-substrate is selected for its high minority carrier electron diffusion length which permits efficient photocurrent collection. Either with a separate diffusion step or by sintering the evaporated or screen-printed aluminum back contact, a p^+ layer is produced at the back side of the substrate. The front contact grid is made of silver, silver-based alloys such as Ag-Ti or Ag-Pd, or multilayer structures like Ti/Pd/Ag, again evaporated or screen-printed. An antireflection coating of SiO_x , TiO_2 or Ta_2O_5 is applied by evaporation or chemical deposition techniques.

The cell current density J is the sum of the photocurrent density and the diode dark current density. In silicon cells at low level injection the contribution by the photocurrent to the forward current is equal to the short circuit current density J_{sc}. We make the further simplification of assuming that only one mechanism, say, diffusion produces the dark forward current. Then we arrive at the most simple expression for the J-V relation for a solar cell.

$$J = J_{sc} - J_0 \left[\exp \left(\frac{qV}{nkT} \right) - 1 \right].$$

J_{sc} is proportional to the photon flux of the incident light. For a fixed spectral distribution of the light, J_{sc} is proportional to the incident power, P_{inc} . By way of J_{sc} the open-circuit voltage V_{oc} also is a function of P_{inc}:

$$V_{oc} \cong \frac{nkT}{q} \ln \left(\frac{J_{sc}}{J_0} \right).$$

The power conversion efficiency of a cell commonly is stated in terms of J_{sc}, V_{oc}, P_{inc} and of the fill factor FF:

$$\eta = \frac{J_{sc} V_{oc} FF}{P_{inc}}$$

The fill factor FF is the ratio of JV, at the maximum power point, to $J_{sc} V_{oc}$.

The reverse saturation current density J_0 is the sum of the diffusion currents from the p- and n-type regions, in our simplified case:

$$J_o = J_n + J_p .$$

With $L_p > W_n$ and $L_n > W_p$ and high rates of surface recombination, conditions typically encountered in silicon cell, J_0 is affected noticeably by heavy doping in the diffused layer. J_p is increased by an effective, higher-than-ideal, intrinsic carrier density n_{ie} :

$$J_p = q / \int_0^{W_n} \frac{N_D(x) dx}{n_{ie}^2(x) D_p(x)} .$$

This non-ideal increase in J_p reduces V_{oc} by about 100 mV below its ideal value to about 600 mV in good laboratory cells. The lowering of V_{oc} by the heavily doped emitter layer is the most prominent detractor from an "ideal" cell efficiency. The physical mechanisms which produce the heavy doping effect still are under discussion. Auger recombination, impurity level broadening and electrostatic reduction of the electron-hole binding energy (and therefore of the energy gap) have been invoked /8,9/. A summary formulation in terms of an effective band gap reduction Δ Eg or of an effective intrinsic carrier density n_{ie} (with $n_{ie} > n_{io}$) has proven adequate:

$$n_{ie}^2 = n_{io}^2 \exp (\Delta Eg/kT).$$

Without heavy doping effects and at constant doping levels the expression for J_0 reduces to

$$J_o = \frac{qn_{io}^2 Dp}{N_D W_n} + \frac{qn_{io}^2 Dn}{N_A W_p}.$$

The p^+ layer at the back contact introduces a p^+p junction which introduces a potential barrier for electrons diffusing toward the back contacts. These electrons are reflected. The electron mirror can raise the short circuit current but more importantly increases V_{oc}. This back surface field (BSF) is particularly desirable for cells with a high base resistivity where many minority carriers may reach the back contact.

Research on Si cells addresses two areas: one, cost reduction of cells for flat panels and two, achievement of very high efficiencies in concentrator cells.

Most of the work on cost reduction addresses problems outside of the domain of device research proper. Included are new and improved methods for the production of Si material, inexpensive fabrication of sheet with good quality, new cell processing techniques and inexpensive contact materials. The passivation of grain boundaries in polycrystalline Si is receiving much attention at present.

The best cell performance so far under terrestrial illumination of 1 kWm^{-2} has been J_{sc} = 35 $mAcm^{-2}$, V_{oc}= 0,625V and FF = 0.80 with η = 17,5 %. The highest V_{oc} has been achieved in an MIS cell using an inversion layer as the n-type emitter and giving V_{oc} = 0.642V. Solar cells for space applications have produced higher currents under under standard test conditions since the Air Mass zero power is 1,4 kWm^{-2}. In 1970, such cells produced J_{sc} ~ 30 $mAcm^{-2}$. The introduction of the "violet" (i.e., violet-sensitive) cell with a very shallow junction raised J_{sc} to 40 $mAcm^{-2}$. Surface texture obtained by etching raised J_{sc} to 45 $mAcm^{-2}$ and the use of a sawtooth cover produced 47,5 $mAcm^{-2}$.

3. SILICON CONCENTRATION CELLS

The high photocurrent densities generated under concentration force
attention to any series resistance which may reduce the fill factor.
Thus, n-type material is employed for most Si concentrator cells,
for the purpose of lowering their internal resistance. It ought
to be noted, however, that the maximum efficiencies predicted for
p^+n and n^+p cells are virtually identical. The performance criteria
of Si concentrator cells are again most conveniently classified by
parameters which affect V_{OC}, J_{SC} and FF, respectively.

The highest efficiency to date, 20.5 % at 70 Suns, has been reached
in a planar p^+nn^+ structure, a refinement simply of the cell for
flat plate applications /10,11/. An efficient concentrator cell is
a successful compromise between a number of conflicting demands.
A thick base (˜ 300 μm) absorbs long wavelength radiation, therefore generating a high photocurrent. A thinner base (˜100 μm or
˜50 μm with a back surface reflector) would be adequate optically
but would also increase the fill factor because of its lower resistance. Unfortunately, 100 μm thick substrates seem to be the
most difficult to handle because of their fragility. So far, thinner
substrates have been made only by chemical etching; successful
direct growth of ˜50 μm substrate materials is considered an important goal for crystal growth applied to both flat plate and concentrator cells. Heavy substrate doping raises the fill factor
through low series resistance; however, light doping increases the
minority carrier diffusion length and thus the photocurrent collection efficiency. A thick, heavily doped emitter is desirable for
a low sheet resistance and thus a high fill factor; by the same
token, more ligth is lost in the "dead layer" of the thick emitter
and heavy doping lowers V_{OC}. The top metal contact grid should cover
as little of the semiconductor as possible for minimum shadowing, yet
the contact area must be large enough for acceptable contact resistance.

Obviously, very high cell efficiencies require concerted tuning and
improvement of many cell parameters. In addition, a large fraction
of Si concentrator research seeks to develop new cell geometries
which may be viewed as variations of the conventional BSF cell /10/.

In the back surface field cell, the base doping level is kept at 10^{16} to 10^{17} cm^{-5}, with the best resistivity ~0.3Ω cm. The p$^+$ emitter is produced by diffusion as well as the n$^+$ back contact region. It is interesting to note that the highest performance is achieved with conventional processing. For instance, ion implantation or rapid annealing techniques have not yet proven of advantage. The n+ region produces a low contact resistance. That it also sets up a back surface field is of little consequence since the hole diffusion length is low in the low-resistance base.

The cell is processed for low base recombination current. In fact, an optimum processing temperature of 300 to 400 C has been postulated /12/. The emitter is so shallow ($^\sim$ 0.3 μm) that few minority carriers recombine while traversing it. Under these circumstances V_{0c} is strongly affected by surface recombination. The back surface field can prevent recombination at the back contact if minority carriers in fact reach it. A similar mirror has been introduced for the emitter (p^{++} p$^+$) which is then called the high-low emitter (HLE). The effectiveness of HLEs is under debate. The high-doping effects in the p^{++} layer may obviate the potential barrier at the p^{++}p$^+$ junction through a reduced effective band gap. A more realistic strategy for the reduction of the recombination rate at the front surface appears to consist in (a) minimizing the metal contact area - with their infinite rate of recombination; an example is the application of contact dots instead of stripes; and (b) passivation of the exposed semiconductor surface, preferentially with a native oxide.

The short circuit current is maximized through optical efficiency and efficient collection of minority carriers. Small contact coverage again is important. A well-matched antireflection coating and surface texturing can reduce reflection losses in the active area to 2 % of the incident light. For concentrator cells the top metal grid is made of silver-based alloys by photolithographic techniques. A high aspect ratio for the fine grid is achieved by electroplating on the previously patterned metal.

The fill factor is made as high as possible by low base resistance, low grid resistance and low sheet resistance of the emitter. The emitter is 0.1 to 0.5 μm deep with a surface concentration in the low 10^{20} cm^{-3} .

At high solar concentration (100 Suns, $J_{SC} \cong 3$ A cm^{-2}) most of the base can be brought into high level injection in the open circuit condition. The base conductance will increase because of the ensuing conductivity modulation in the entire base expect in the vicinity of the back contact. However, at some forward bias, lower than V_{OC} but possibly including the operating voltage, a wider region near the base contact loses conductivity modulation. Then the series resistance will increase. This increase is less severe for n-type than for p-type bases. Therefore, n-type bases have been chosen for most concentrator cells. The high level injection condition in the base also increases the effective carrier lifetime /13/. This effect causes J_{SC} to increase more than linearly with increasing light concentration factor /14/. It is also possible that high level injection could lead to effective band gap narrowing analogous to the high doping effect discussed earlier.

4. CONCENTRATION CELLS WITH COMPOUND SEMICONDUCTORS

The absorption length for light with higher than band gap energy is only ˜1 μm for GaAs and InP. Such small absorption length qualifies these and related III-V semiconductors as well as their alloys for thin film cells. In single crystal form, i.e., with a minimum of defects, III-V semiconductors can be made into very efficient solar cells for concentrator applications. To achieve high V_{OC} , the active layer is grown epitaxially. GaAs cells fabricated by liquid phase epitaxy on single crystal GaAs substrate wafers have reached the highest efficiency - 24,7 % at 178 Suns - of any single cell so far /15/. The key to obtaining high photocurrent in such cells where the light is absorbed very near the surface, is good surface passivation. In the conventional GaAs cell the surface of a p-on-n GaAs diode is passivated with a p-type alloy of the lattice-matching system $(Al_xGa_{1-x})As$. The p-GaAs layer and thus the junction is formed when the Zn-doped $(Al_xGA_{1-x})As$ is grown: Zn diffuses into the n-GaAs. The same type of cell has been fabricated by metal-organic chemical vapor deposition, a process better suited to high production rates than LPE /16/. The GaAs-based buried homodiode or heterostructure homodiode cell is the type of cell structure which is considered the basic building block for multijunction cells with very high efficiency.

Efficient n- on p homodiode cells have been fabricated by halide chemical vapor deposition of epitacial GaAs or InP layers on the respective substrate materials. These diodes are surface passivated with anodic oxides. To ensure absorption of most of the incident light below the np junction the n-type layers are less than 0.1 μm thick /17/.

Highly efficient cells, with a single junction, of compound semiconductors are also of interest for space applications because of their radiation hardness. Very thin cells are desirable for a high power/weight ratio /18,19/. For terrestrial use they may face stiff competition from silicon concentrator cells. However, compound semiconductors are indispensable to the multijunction cell concepts which we shall discuss next.

5. CASCADE CONCENTRATORS

Cascade, or multijunction, or tandem - when there are two junctions - cells convert different portions of the solar spectrum in separate devices. The solar spectrum is split in two or more aliquot portions and it is converted in cells with the appropriate energy gaps. In this way, the degradation of the energy of an absorbed photon to the value of a single band gap is avoided. An effeciency of over 30 % has been predicted for tandem cells, and more than 35 % for higher multijunction cells.

The most efficient cascade cell to date was a tandem of two separately mounted cells illuminated by way of a dichroic, or spectrum-splitting, filter. The filter transmitted light with $h\nu >$ 1.6 eV onto a cell based on $Al_{0.2}Ga_{0.8}As$, and reflected lower energy light onto a silicon cell. $Al_{0.2}Ga_{0.8}As$ (Eg = 1,65 eV) and Si (Eg = 1.11 eV) are not optimally matched, yet a conversion efficiency of 27 % was achieved at 160 Suns /20/. The highest tandem cell efficiency is expected for a combination of a 1.0 eV with a 1.7 eV band gap cell. The most compact version of a tandem cell is monolithic. The two - or more - cells are fabricated on top of each other. III-V compound semiconductors are employed for such devices. They can provide a wide range of energy gaps and offer the possibility of single crystal growth by the matching of lattice parameters. In a monolithic tandem cell, with the large gap diode on top of the small gap diode, the cells are in series both optically and electrically. The voltages of the individual cells add, while the total current is the smaller of the two individual currents. There-

fore, a careful match between the band gaps is essential. By the
same token, any change in the spectral distribution of the incom-
ing light may cause a drop in cell efficiency.

The highest efficiency with a monolithic tandem cell so far (16,4 %
at 1 Sun of AM1) was obtained with (Al,Ga)As alloy devices /21/.
The (Al,Ga)As system cannot provide the optimum combination of
band gaps but offers a mature materials technology. Although this
cell is more simple than tandem cells made with quarternary alloys,
the complexity of layers (six layers on a GaAs substrate) reminds
one of modern injection lasers.

Two alloy systems can provide the desired band gap match: (Al,Ga)
(As,Sb) and (Al,Ga,In)As. Tandem cells may be made with a single
lattice parameter throughout the multilayer structure for the desired band gap ranges of 0.95 to 1.20 eV and 1.65 to 1.80 eV. However, there is a shortage of substrate materials (Ge on Si, GaAs,
InP) if one does not resort to single crystal alloys. The lattice
parameter of the substrate may have to be adjusted by the growth
of graded alloys or of multilayer structures.

Another critical aspect of monolithic multijunction cells is the
need for low-resistance connecting junctions between the individual
cells. These junctions must form a contact between an n- and a p-
layer with low resistance and low optical absorption loss. Tunnel
junctions are the most obvious choice for this task. Concern about
the long term stability of such tunnel junctions remains to be
alleviated. The difficulty of including a tunnel junction while paying attention to the other constraints mentioned earlier have led
to investigations of other connection techniques /22/. Separate
cells series-connected by wire, or separately wired (three- or
four-terminal) arrangements have been proposed.

Cascade cells promise very high conversion efficiencies. While the
concept of voltage addition has been proven, no truly competetive
cascade cells have emerged yet because of the formidable demands
on materials technology which they pose. The development of cascade cells is likely to track that of other optoelectronic and
logic devices based on III-V semiconductors.

6. THE THERMOPHOTOVOLTAIC CELL

The thermophotovoltaic (TPV) cell is a concentrator cell not optimized for solar radiation but for the radiation of a black body heated to a lower temperature (2000 to 2200 K instead of 6000 K) /23/. The black body ultimately is to be heated with concentrated sunlight. The radiator and the TPV cell form an optical cavity. Radiation with $h\nu > E_g$ is absorbed and converted by the cell. Below-band gap-radiation is transmitted through the cell, reflected at the back surface of the cell and returned to the cavity for recycling.

To date, the TPV cell has been realized with silicon technology. A laboratory setup with an electrically heated radiator has reached 30 % conversion efficiency. The most critical requirement for a TPV converter appears to be a radiator material which is stable against evaporation followed by redeposition on the surface of the cell. This degradation mechanism reduces the efficiency for optical recycling. In addition to the usual demands placed on a concentrator cell, the TPV cell needs: A highly reflective back surface, realized with a small contact area and a silver reflector; a low surface recombination rate, realized with a passivating thermal oxide, particularly between the silver mirror and the silicon; and minimum below-band gap optical absorption in the bulk.

7. THIN FILM CELLS

Thin film cells are geared to low production cost through the consumption of a minimum of material per unit aperture area, and through high production rates. A thin film does not exclude high efficiency. It is a common premise that high production rates do. Therefore a continuing debate is going on about the lowest acceptable efficiency for thin film cells. Small markets will be satisfied with low efficiency. A good example is the amorphous hydrogenated silicon cell. Devices with a few square centimeters are being incorporated in watches and pocket calculators. These cells, with $\eta \approx 5$ % in daylight, are about twice as efficient under fluorescent lamps because of a better spectral match. Amorphous thin film cells satisfy the small appliance market well, but the principal target of thin film cell development is power generation on a large scale. Utility-like power generation will require a module efficiency or 10 % or possibly 15 % as mentioned in the introduction. Until this year the research on thin film cells has been characterized by a

profusion of semiconductor materials, elementary as well as
sophisticated fabrication technology, a variety of device types
and most of all, a push to exceed the magic efficiency number of
10 %. Four thin film cells now have surpassed this value:

$p-Cu_2S/n-/ZN,Cd)S$ in 1981 /24/;
$p-CuInSe_2/n-(Zn,Cd)S$ in 1982 /25,26/
$p-CdTe/n-CdS$ in 1982 /27/;
$p-\alpha SiC_\gamma H_x/i-\alpha SiH_x /n-\alpha SiH_x$ in 1982 /28/.

The first three cells are polycrystalline and at least formally
analoga of cells that may be made in single crystal form. The
fourth cell is made of amorphous semiconductors. We shall bring
up these two types of thin film cells in the same sequence. Before
we do so, we ought to discuss the typical philosophy of thin film
cell research and development.

Any semiconductor with a band gap between ~ 1.0 eV and ~ 1.7 eV has
the potential for a high photovoltaic conversion efficiency.
Virtually all known semiconductors in this range have been considered for photovoltaic (probably many more materials exist that
fit the energy gap but have not been characterized). After either
advancing nor neglecting the argument whether raw materials are
available in sufficient quantity, first efforts concentrate on demonstrating a photovoltaic effect. Next comes the fabrication of
a device with ~ 5 % efficiency. This has proven fairly easy with a
number of semiconductors both in the form of single crystals and
of polycrystalline films. Other did not reach even 5 % and were
terminated. Beyond such initial study, the principal research goal
for a small group is the attainment of 10 % efficiency. Around
this value the improvement of laboratory efficiency becomes so
difficult that it has to be carried by a larger group. Before such
commitment is made, the following criteria must be applied to the
cell at hand.

- Efficiency. Improvement of the laboratory efficiency remains an
important task. Around $\eta = 10$ %, progress becomes slow because
usually neither the device physics nor the materials are well understood. Pushing up the efficiency is therefore a highly empirical
undertaking. It is best accompanied by a broad research program on
these aspects.

- Cell area. Often the active area of early research cells is only a few mm^2. It is a common experience that the efficiency drops as the cell area is increased. The drops seem to be caused by series resistance (which lowers the fill factor) or by the inclusion of gross defects (which lower V_{oc}) of low area density. For research cells an area of 1 cm^2 is considered respectable. Early attempt to increase the cell size are made to test fabrication technology, in the course of experiments on device geometry, or with a marketable product in mind.

- Stability. A cell which approaches η = 10 % with a size of ~1 cm^2 will be critically evaluated as a candidate for commercial development. The next important information demanded from the laboratory is data about the long term (up to 30 years) stability of the cell when incorporated in module. Obviously it is impossible to provide a reliable answer. The mechanism of degradation, if any is observed, usually is not understood so that it is impossible to specify reliable procedures for accelerated life testing. Confusing results may be obtained in early stability tests. Cells may not fail by the most expected mechanism. Cells may fail at room but not elevated temperature. Failure may be irreproducible. Usually the fabrication of a few stable cells, out of al larger number, is taken as sort of proof of feasibility for stable cells. As a consequence of the confusing stability situation, thin film modules have been introduced prematurely to the market. Other modules are not yet marketed for fear of failure in the field.

- Availability of materials. The availability of the constituent chemical element is an important criterion in the selection of a cell for technology development. Often the cost reflects the availability. The availability has been gauged by comparing the potential annual consumption for photovoltaic manufacture with the current annual production. A more encompassing yardstick of availability is the module area that could be produced by using up all of the world's estimated resources of a scarce element!

- Toxicity of materials. This criterion is difficult to evaluate, particularly for a technologist. It is bound to play a role once a large industry begins to emerge.

- Process scalability and manufacturing technology. Research results which encourage further cell development may have been obtained

with a process that cannot be scaled to large area and throughput.
An example is the co-evaporation of several chemical elements to
form a compound semiconductor. In such a case processes must be
developed which can be scaled up in size and in rate. The technology for these processes should already have been tried for other
applications.

We will discuss now the status of thin film cell research in two
sections. One comprises the polycrystalline thin films. The other
deals with amorphous silicon. Polycrystalline cells are derived
from single crystal analoga, at least in principle, so that our
classification is plausible. However, different classifications
are conceivable. For instance, one might form one group of the
thin film cells developed after their single crystal analoga, and
the other group of thin film cells discoverd as such (Cu_2S/CdS,
amorphous silicon).

8. POLYCRYSTALLINE THIN FILM CELLS

We mentioned earlier that it is not difficult to make a 5 %
efficient cell with the proper semiconductor, particularly if
it is a single crystal. Si, Ge, GaP, GaAs, InP, CdSe, CdTe, InSe,
WSe_2, and $CuInSe_2$ reached 5 % in brief laboratory studies.
η = 10 % often can be achieved with an extended effort but without serious difficulty. Si, GaAs, InP, CdTe, $CuInSe_2$ are good
examples. One must be able to draw from a body of device technology to bring a single crystal cell to $\eta \geq$ 15 %. Highly efficient
Si, GaAs and InP cells had to rely on know-how developed for other
device applications.

All of the 5 % materials have been considered for polycrystalline
thin film cells. The first extra condition put on the semiconductor is a short optical absorption length ($1/\alpha \cong$ 1 μm). This condition rules out semiconductors without a direct lowest band gap,
or at least a direct gap with an energy close to any lowest in-
direct gap.

The existence proof of a single crystal cell with a high absorption coefficient is encouraging. Still, it has proven impossible so far to predict the success of a program geared to
the preparation of a 10 % efficient laboratory specimen of a
thin film cell.

Qualitatively the causes for this unpredictability are understood. The electronic effects of a high density of grain
boundaries (grain size ~ 1 µm) can neither be predicted nor
controlled. An accepted guideline is that the effects of grain
boundaries become less detrimental as one proceeds from the
group IV via the III-V to the II-VI semiconductors. The concept
of a thin film cells also implies a high density of interfaces.
Layer thicknesses of ~1 µm are typical. These interface often
join dissimilar or even reactive materials. Often there is
greater flexibility in fabricating or controlling the interfaces in the single crystal counterparts. For example, a heavy
back contact diffision is common in single crystal devices.
Such treatment is unacceptable for an active layer whose
thickness is ~ 1 µm. In fact, many thin film cell research
programs are plagued by excessive reaction of the semiconductor with the metal of the back contact. The successive
layer-by-layer buildup so desirable for eventual cell manufacture
proves a curse to cell research. Each subsequent processing step
may change the properties of any layer or interface fabricated
earlier. Of course this problem is common to all multilayer
device technologies such as integrated circuits or optoelectronics. It is aggravated by the comparative lack of knowledge
about many thin film photovoltaic materials. As an illustration
we again invoke the ohmic contact to a p-type absorber layer.
This layer is deposited on the ohmic contact material. An
inert material and a low deposition temperature are desirable.
A conflict arises with the demand for relatively large and de-
should be nucleated and grown at high substrate temperature.

These complications of course obtain also with the cells which are
not analoga of single crystal devices. They have had two con-
sequences for thin R&D programs. The programs have perforce
become quite comprehensive. They cover semiconductor physics and
characterization, deposition techniques, device technology,
measurement and modeling. Furthermore, the requisite size and cost
of the research projects have been prompting persistent review
and questioning of what ought to be long range programs.

Thin-film cells may be defined as those whose absorber layer is
a few micrometers thick. Such definition excludes silicon which

requires 50 μm to absorb 90 % of the above-band gap light. On the other hand, if all films deposited on substrates are included in the category of thin films, some silicon devices qualify. In the latter case, a large lateral grain size is important. The substrate for thin film silicon cells usually is an impure grade of coarsely polycrystalline silicon. Carbon sheets also have been used. η has reached ~ 10 % /29,30/.

The III-V semiconductors have been disappointing in thin film cells. The highest reported efficiencies are 8,5 % for a n-GaAs/Ag Schottky barrier cell and 5,7 % for a p-InP/n-CdS heterodiode, each about 0,4 of the efficiency of the single crystal counterpart /31/. While J_{SC} and FF may be somewhat lower than in single crystal devices, the main detractor has been V_{OC} . In general it has proven easier to achieve satisfactory J_{SC} than V_{OC} in polycrystalline cells. In view of the dicussion of thin film cells of about five years ago this finding comes as a surprise. The well understood grain size criterion for J_{SC} (size > $1/\alpha$ $\cong$ L) had received much attention while the grain boundary effects on V_{OC} were not understood and little discussed. In any case, J_{SC} in p-type materials ($L_n \cong 1/\alpha$) has come close to the theoretical value. Not only is $1/\alpha$ larger than the grain size, but also a sizable fraction of the incident light is absorbed in the diode space charge region. The low V_{OC} appears to result from non-ideal interface within or near the junction space charge; and from cell shunting by highly conducting grain boundaries. A high V_{OC} has been measured on individual grains of p-GaAs/n-AlAs polycrystalline cells. However, the n-AlAs is laterally insulting, due to high grain boundary resistance, so that the photocurrent cannot be collected.

Two II-VI compound semiconductors exhibit the proper band gap: CdSe (1.71 eV) and CdTe (1,44 eV) /32/. CdSe can be made only n-type. It has found some interesting research applications in photoelectrochemical anodes but is widely discounted as serious contender for solid state photovoltaics. Of an exclusively n-type semiconductor, no pn junctions can be made; no adequate p-type large gap window material exists for the fabrication of a heterodiode; and Schottky barriers with semitransparent (100 to 400 $\overset{o}{A}$ thick) metal layers are considered unreliable.

CdTe can be made both p- and n-type. The CdTe homodiode requires
a shallow junction because of the short optical absorption length
over the solar spectrum. The cell suffers from the usual short-
comings of non-passivated homodiodes in direct gap semiconductors:
excessive surface recombination and high sheet resistance. Furthermore, minority carrier diffusion lengths in doped CdTe often are
low so that current collection from the bulk can be unsatisfactory.
The p-CdTe/n-CdS heterodiode does away with several of the drawbacks of the unpassivated homodiode, at the expense of some loss
in usable photon flux. This cell is made either as a p-CdTe/n-CdSe
heterodiode or as a CdS-passivated np-CdTe homodiode. (There are
no reliable means for the identification of very shallow - of the
order of 0.1 μm - np junctions). Single crystal cells have reached η = 12 %, thin film cells 10 % /27/, and screen-printed "thick
film" cells 8 % /33,34/.

CuInSe$_2$, a chalcopyrite-type semiconductor, is derived from the
II-VI semiconductors. Its ban gap is 1,01 eV, and it can be made
both n-and p-type. Single crystal p-CuInSe$_2$/n-CdS heterodiode cells
have reached η = 12 %. The thin film device p-CuInSe$_2$/ n-(Zn,Cd) S
has attained 10,6 % /26/ with J_{sc} equal to the theoretically possible value. The optical absorption coefficient above band gap
is $\sim 10^5$ cm^{-1} , so that most of the light is absorbed in the
junction space charge. Under simple test conditions (80 ^{0} C,
ambient air, solar simulator illumination) this device is surprisingly stable, particularly in view of the known tendency of
copper to diffuse.

p-Cu$_2$S/n-CdS is the original thin film cell. p-Cu$_2$S/n-(Zn, Cd)S
was the first thin film cell to reach η = 10 % /24/. This cell
was transferred to production prematurely, without paying heed
to instability whose causes were not well understood. Two companies in the U.S. have suspended their plans for production while
continuing research. A new production facility in being set up in
Germany.

9. AMORPHOUS HYDROGENATED SILICON

α-SiH$_x$ appears to meet many of the traditional criteria for a
thin-film photovoltaic material. Above its optical absorption
edge (1.70 to 1.85 eV or 0.73 to 0.67 μm) it absorbs light strongly ($\alpha > 10^4$ cm^{-1}) so that a layer of a few μm suffices. Amorphous

alloys with lower (addition of Ge or Sn) or higher (addition of C or N) gaps can be prepared for the fabrication of cascade cells. α-SiH$_x$ can be deposited at substrate temperatures of 250 to 400 C; low substrate temperatures imply the use of inexpensive substrates and the easy development of continuous processing. The component materials for the semiconductor itself are abundant. The amorphous structure implies the absence of grain boundaries and their detrimental effects on thin film cells. Furthermore, an amorphous structure typically is associated with a very high rate of production - in the sense that when a material forms by nucleation and growth, the size of structural features is inversely proportional to the rate of formation. Because α-SiH$_x$ is a new type of semiconductor for pn junction devices, it has elicited tremendous interest in the research community. Many groups are working to understand its basic physical properties. This body of information produces valuable feedback to solar cell development.

However, α-SiH$_x$ also shows some serious disadvantages. For one, the prediction of its maximum practical solar conversion efficiency has been entirely empirical and has trailed cell development. This prediction, based on the performance of laboratory single junction device, now stands at 12,5 % for AM1. The predicted efficiency for a tandem junction devices is 21 % and for a triple junction device, 24 % /35/. The maximum single cell efficiency is low largely on account of the high energy of the optical absorption edge. The long-term stability of the cell is questionable; cell efficiency degradation has been observed both in the dark (e.g., due to chemical reaction between metal contacts and α-SiH$_x$) and under illumination (charge trapping and possibly photon-induced restructuring). The rate of film growth is 1 to 5 Å per second, so that the time required to fabricate a cell is ~1 hour.

The highest cell efficiency reported so far is 10 % /28/. The structure of this cell is as follows. A transparent conductor (indium tin oxide or tin oxide) is deposited on a glass plate. A boron-doped α-SiC$_\gamma$H$_x$ layer (~ 7 nm) follows. The boron doping reduces the optical transmittance of α-SiH$_x$. Carbon is introduced to shift the absorption edge to higher energy in order to increase the transmittance. The undoped semi-insulting layer of α-SiH$_x$ follows (0.5 to 1.0 µm). This i-layer is the active absorber since it has adequate minority carrier (or ambipolar) diffusion lengths L. L up to 1,3 µm has been measured. The electric field in the device is established by deposition of a 10 nm thick phosphorus-doped α-SiH$_x$ layer. Contact to this layer is made with titanium.

Since α-SiH$_x$ represents a new class of semiconductors, ample research on its physical properties /36/ and device applications /37/ is being published.

10. SUMMARY

We have presented a brief discussion of the research on prominent photovoltaic devices, together with some criteria for device research. It is too early to tell whether photovoltaics will come to produce power on a large scale. The necessary performance and and price certainly are within the capability of known semiconductors, at least in principle. However, a successful solar cell must meet an extensive list of requirements. It will take a long period of research and development to single out this device.

11. ACKNOWLEDGEMENTS

Much of this paper was written during a stay at the Physics Department of the University of Constance. I would like to thank Professor Ernst Bucher for his hospitality. This work was supported by the Solar Energy Institute under Subcontract No. XL-2-02075-01.

REFERENCES

1. For the term "mass electricity" I am indebted to Dr. Elliott Berman of ARCO Solar.

2. R.W. Taylor, "Utility Requirements for Photovoltaic Power", Abstract No. 40, Symposium on Materials and Processing Technologies for Photovoltaics, Electrochemical Society Spring Meeting, Montreal, May 9-14, 1982.

3. H.J. Hovel, "Solar Cells," Vol. 11 of "Semiconductors and Semimetals," R.K. Willardson and A.C. Beers, editors, Academic Press, New York 1975.

4. M.A. Green, "Solar Cells," Prentice-Hall, Englewood Cliffs, N.J., 1982.

5. S.J. Fonash, "Solar Cell Device Physics," Academic Press, New York 1981.

6. R.H. Bube and A. Fahrenbruch, "Solar Cell Materials," Academic Press, New York 1982.

7. Conference Records of the IEEE Photovoltaic Specialists Conferences, 1st (1961) to 16th (1982) Conference. Abbreviated to PVSC in later references.

8. H.P.D. Lanyon, Solar Cells 3 (1981) 289.

9. D. Redfield, Solar Cells 3 (1981) 313.

10. R.J. Schwartz, Solar Cells 6 (1982) 17.

11. R.D. Nasby, C.M. Garner, F.W. Sexton, J.L. Rodriquez, B.H. Rose and H.T. Weaver, Solar Cells 6 (1982) 49.

12. J.G. Fossum and D.S. Lee, 15th PVSC (1981) 120.

13. V.L. Dalal and A.R. Moore, J. App. Phys. 48 (1977) 1244.

14. R.D. Nasby, C.M. Garner, H.T. Weaver, F.W. Sexton and J.L. Rodriguez, 15th PVSC (1981) 132.

15. R. Sahai, D.D. Edwall and J.S. Harris, Jr., Appl. Phys. Lett. 34 (1979) 147.

16. P.E. Gregory, P.G. Boden, M.J. Ludowise, C.B. Cooper III and J.R. Saxena, 15th PVSC (1981) 147.

17. G.W. Turner, J.C.C. Fan, R.L. Chapman and R.P. Gale, 15th PVSC (1981) 151.

18. J.C.C. Fan, C.O. Bozler and R.W. McClelland, 15th PVSC (1981) 660.

19. R.J. Boettcher and P.G. Borden, 16th PVSC (1982), to be published.

20. L.W. James, H.A. Vander Plas and R.L. Moon, Final Report, Sandia contract 07-6953 (1979).

21. S.M. Bedair, J.A. Hutchby, J. Chiang, M. Simons and J.R. Hauser, 15th PVSC (1981) 21.

22. J.C.C. Fan and B.-Y. Tsaur, 16th PVSC (1982), to be published.

23. R.W. Swanson, Technical Digest of the 1980 IEDM, Washington, D.C. 8-10, 1980. IEEE, New York 1980; p. 186.

24. R.B. Hall, R.W. Birkmire, J.E. Philips and J.D. Meakin, 15th PVSC (1981) 777.

25. R.A. Michelsen and Wen S. Chen. Abstracts of the Polycrystalline Thin Film Review Meeting, SERI, Golden, CO, May 3-5, 1982, p. 17.

26. R.A. Mickelsen und W.S. Chen, 16th PVSC (1982), to be published.

27. Y.S. Tyan and E.A. Perez-Albuerne, 16th PVSC (1982), to be published.

28. D.E. Carlson, U.S.-Japan Joint Seminar on "Technological Applications of Tetrahedral Amorphous Silicon, "Palo Alto, CA, July 19-23, 1982.

29. T.L. Chu, S.S. Chu, C.L. Lin and R. Abderrassoul, J. Appl. Phys. 50 (1979) 919.

30. C.P. Khattak, M. Basaran, F. Schmid, R.V. D'Aiello, P.H. Robinson and A.H. Firester, 15th PVSC (1981) 1432.

31. D.L. Feucht, 15th PVSC (1981) 648.

32. S. Wagner, 16th PVSC (1982), to be published.

33. N. Nakayamo, H. Matsumoto, A. Nakano, S. Ikegami, H. Uda and T. Yamashita, Japan. J. Appl. Phys. 9 (1980) 703.

34. H. Uda, H. Matsumoto, Y. Komatsu, A. Nakano and S. Ikegami, 16th PVSC (1982), to be published.

35. Y. Tawada, K. Tsuge, M. Kondo, H. Okamoto and Y. Hamakawa, 16th PVSC (1982), to be published.

36. R.A. Street and D. Biegelsen, "Physics of Tetrahedrally Coordinated Amorphous Semiconductors," American Physical Society, New York 1982.

37. Proceedings of the 3rd Photovoltaic Science and Engineering Conference in Japan, Kyoto, May 19-21, 1982. Japan J. Phys., Supplement 21-2 (1982), to be published.

Advanced Device Technology for High Speed GaAs VLSI

Masayuki Abe, Takashi Mimura, Naoki Yokoyama,
and Katsuhiko Suyama

Fujitsu Laboratories Ltd., Fujitsu Limited,
1015 Kamikodanaka, Nakahara-ku, Kawasaki 211, Japan

SUMMARY

Recent advances in device technology for high speed GaAs VLSI, that
is, self-aligned fully-implanted GaAs MESFET technology and High
Electron Mobility Transistor (HEMT) technology, are reviewed. Basic
GaAs IC devices and circuits and the current status of GaAs MSI/LSI
devices are also summarized here from the viewpoint of practically
available technology. Using self-alignment technology, we first
demonstrated MSI/LSI complexity with a DCL, 4x4-bit and 6x6-bit
parallel multipliers, and a 1K-bit sRAM cell array. The HEMT is a very
promising device for VLSI because of its ultra-high-speed performance
and its low power dissipation at liquid nitrogen temperature. E/D type
Direct Coupled HEMT logic using a 1.7 μm gate length achieved a
switching time of 17.1 ps at the liquid nitrogen temperature and, more
recently, a switching time of 12.8 ps using 1.1 μm gate HEMT logic,
which is the highest-reported speed for any device logic. Furthermore,
projected system performances for future large scale computer and the
prospects of high speed GaAs VLSI will be discussed.

1. INTRODUCTION

The evolution of high-speed GaAs integrated circuits (ICs) is the
result of continuous technological progress utilizing the superior

electronic properties of GaAs. Since the first GaAs ICs were developed
by Van Tuyl et al. in 1974 (1), device, circuit design, processing, and
material technologies have progressed and continue to grow rapidly.
Figure 1 shows the evolution of GaAs ICs. The complexity of GaAs IC
evolution projected by Nuzillat (2) has led to a growth rate that
increases threefold each year. GaAs MESFET MSI/LSI complexities have
already been achieved with Buffered FET Logic (BFL) (Van Tuyl el al.
(3), Suyama et al. (4)). The largest BFL circuit, a 4-bit Arithmetic and
Logic Unit (ALU) with a high-speed digital processing time of 2 ns
despite 2 μm design rule technology, contains 99 gates (Suyama et al.
(4)). Schottky Diode FET Logic (SDFL) circuits (Eden et al. (5), Long
et al. (6), Lee et al. (7), (8)), using depletion-mode FETs, have
achieved high-speed performances compared with Si LSI. The largest
integration of SDFL circuits, an 8x8-bit binary multiplier with a
multiplying time of 5.3 ns, contains 1008 gates (Lee et al. (8)).
Future high-speed LSI/VLSI complexity can be achieved with the Direct
Coupled Logic (DCL) circuit using enhancement-mode FETs because of
extreme circuit simplicity and low power dissipation. The DCL circuit,
developed by Ishikawa et al. (9) in 1977, is shown in Fig. 1. The
complexity of enhancement-mode circuits was limited until such
breakthroughs as ion implantation technology achieved uniform control
of GaAs IC fabrication.

   Two further, important breakthroughs have been made in high-speed
GaAs VLSI technology: (1) in self-alignment device technology to obtain
reproducible characteristics of enhancement-mode MESFETs (Yokoyama et
al. (10), (11), Suyama et
al. (12), Nakayama et al.
(13)), and (2) in High
Electron Mobility
Transistor (HEMT)
technology based on
Molecular Beam Epitaxy
(MBE) technology enabling
the fabrication of
modulation doped
super-mobility
heterojunction devices
(Mimura et al. (14), (15),
Abe et al. (16), Tung et
al. (17)). As Fig. 1
illustrates, complexity
has increased for GaAs
VLSI using DCL.

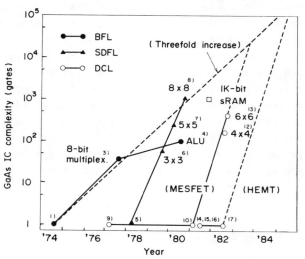

Fig. 1. Evolution of GaAs IC complexity.

We will first present GaAs IC devices, circuit approaches, and
current status of GaAs MSI/LSI devices using practically-available
technology. Next, we will describe recent advances in GaAs MESFET IC
technology, a self-aligned GaAs MESFET using high temperature stable
TiW silicide gate technology, and demonstrate LSI logic and memory. We
will then review HEMT technology using modulation doped GaAs/AlGaAs
heterojunction structure grown by MBE and discuss projected system
performances for future large scale computer, and prospects for
high-speed GaAs IC technology.

2. GaAs IC DEVICES AND CIRCUIT TECHNOLOGY

2.1 Basic logic devices and circuit approaches

Many new devices such as FET or bipolar devices have been proposed
and/or developed over the last few years. Bipolar-like devices include
bipolar transistors, heterojunction bipolar transistors (Kroemer (20)),
and hot electron transistors
(Heiblum (21)). FET devices include
MOSFETs (Yokoyama et al. (22)),
MESFETs (Eden et al. (23)), JFETs
(Zuleeg et al. (24)), heterojunction
FET devices (HEMT) (Mimura et al.
(14)), and permeable base
transistors (Bozler et al. (25)).
These devices are reviewed in Eden
(18) and Solomon (19).

We will confine ourselves here to
GaAs logic devices for LSI/VLSI,
i.e., MESFETs, JFETs, and HEMTs.
Figure 2 compares schematic cross
sections of the first planar
MESFETs, self-aligned MESFETs,
JFETs, and HEMTs. Figure 2 (a)
shows the planar D-MESFET fabricated
(Eden (18)) by ion implantation into
a semi-insulating GaAs substrate.
The basic structure is very simple,
consisting of a thin (1000 Å)
n-type (typically 2×10^{17} cm^{-3})
active region joining two ohmic
contact with a narrow metal Schottky

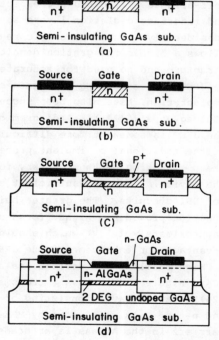

Fig. 2. GaAs IC device structure,
(a) planar MESFET, (b) Self-aligned
MESFET, (c) JFET, and (d) HEMT.

barrier gate (1 μm long) separating the source and the drain. The
D-MESFETs are very widely used, and circuits employing D-MESFETs pose
the least problem as far as fabrication is concerned. Any region of
the source-drain channel not under the gate is conductive and precise
gate alignments are not necessary, nor are special gate recesses or
other means for avoiding parasitic source and drain resistance.
Enhancement-mode MESFETs have simple circuits because the logic gates
require only one power supply as opposed to the two power supplies
needed for D-MESFET circuits.

Figure 2 (b) shows a self-aligned MESFET. In conventional MESFETs,
extension of the surface depletion layer cannot be avoided because of
traps localized at the GaAs surface, source series resistance is also
very high because of the thinness of the undepleted n layer. Extension
of the interface depletion layer because of traps near the interface
between the active layer and the substrate is also a factor which may
significantly increase drain series resistance. High source and drain
series resistance (parasitic resistance) and fluctuations in the
charges of these traps adversely affect the performance and
reproducibility of such FETs. In self-aligned MESFETs, however,
self-aligned n^+ regions are expected to prevent extension of surface
and interface depletion layers so that the undepleted n^+ layer
considerably reduces parasitic resistance. Self-aligned MESFETs also
allows a higher integration density because of their full planar
structure and no need for accurate gate alignment.

JFET devices, using a p^+ gate stripe formed by selective ion
implantation, have also been successfully employed in GaAs digital ICs
(Zuleeg et al. (24)). A diagram of a GaAs JFET is given in Fig. 2.
The JFET is somewhat more difficult to fabricate than a MESFET because
of the additional p^+ implant process steps and the precise control of
the p^+ junction depth necessary to control the pinch-off voltage of
the device. However, sufficient control has been obtained, at least
for SSI circuits. The greater built-in potential of the p^+-n
junction provides a higher forward-bias gate conduction limit
(approximately 1.1 V) which should provide a significant speed
advantage for enhancement-mode JFETs over MESFETs.

High mobility FET devices are also being developed for use in GaAs
ICs. Figure 2 (d) shows a cross section of a HEMT (Mimura et al.
(26)). An epilayer consisting of undoped GaAs, Si doped n-type AlGaAs,
and n-type GaAs, are grown by MBE on a semi-insulating GaAs substrate.
Carriers in the AlGaAs layer transfer adjacent undoped material by
virtue of the differences in electron affinity between the two (Dingle
et al. (27)). These devices use to good advantage of the greatly
reduced phonon scattering in a high purity undoped n-type GaAs channel

Table 1. Comparison of basic logic gate approaches.

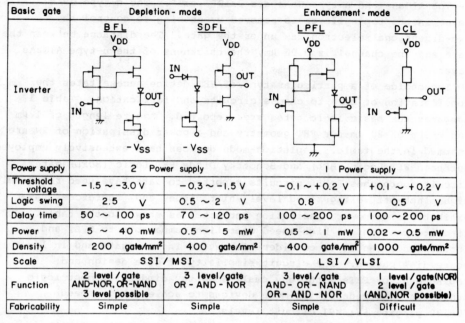

Basic gate	Depletion - mode		Enhancement - mode	
	BFL	**SDFL**	**LPFL**	**DCL**
Inverter				
Power supply	2　Power supply		I　Power supply	
Threshold voltage	−1.5 ～ −3.0 V	−0.3 ～ −1.5 V	−0.1 ～ +0.2 V	+0.1 ～ +0.2 V
Logic swing	2.5　V	0.5 ～ 2　V	0.8　V	0.5　V
Delay time	50 ～ 100 ps	70 ～ 120 ps	100 ～ 200 ps	100 ～ 200 ps
Power	5 ～ 40 mW	0.5 ～ 5 mW	0.5 ～ 1 mW	0.02 ～ 0.5 mW
Density	200 gate/mm²	400 gate/mm²	400 gate/mm²	1000 gate/mm²
Scale	SSI / MSI		LSI / VLSI	
Function	2 level / gate AND-NOR, OR-NAND 3 level possible	3 level / gate OR - AND - NOR	3 level / gate AND- OR - NAND OR - AND - NOR	1 level / gate(NOR) 2 level / gate (AND, NOR possible)
Fabricability	Simple	Simple	Simple	Difficult

Table 2. Ring oscillators speed-power performance for
GaAs IC device approaches.

Source		Approach	Gate length & width (μm x μm)	Switching delay (ps)	Speed-power product (fJ)	Fan-in / Fan-out
Hughes	(29)	D - MES/BFL	0.5 x 50	34	1400	1/1
	(30)	E - MES/DCL	0.8 x 25	25	80	1/1
HP	(3)	D - MES/BFL	1.0 x 20	86	3900	2/2
Rockwell	(32)	D - MES/SDFL	1.0 x 10	52	63	2/1
Fujitsu	(33)	E - MES/DCL	1.2 x 20	170	120	1/1
	(10)	E - MES/DCL (Self-aligned)	1.5 x 30	50	287	1/1
	(15)	E - HEMT/DCL	1.7 x 7	56	26	1/1
		(77K)	1.7 x 13	17	16	1/1
NTT	(34)	E - MES/DCL	0.6 x 20	30	57	1/1
		(77K)	0.6 x 20	18	616	1/1
Thomson CSF	(35)	D - MES/BFL	0.75 x 20	68	2000	1/1
	(36)	E - MES/LPFL	1 x 35	105	230	1/1
	(37)	TEGFET/DCL	0.7 x 20	18	17	1/1
Mc Douglas	(38)	E - JFET/DCL	1.0 x 10	150	60	1/1

at the temperature of liquid nitrogen. The doping concentrations in n-type AlGaAs and n-type GaAs are $2x10^{18}$ cm^{-3}. The n-type GaAs layer in the top epilayer is removed by etching to control the two-dimensional electron gas under the gate. The distance between the gate and the channel is 0.06 μm, the thickness of the n-type AlGaAs layer.

The choice of a particular type of FET device necessitates the consideration of here to choose circuits and fabrication. Table 1 compares the basic logic gates ever reported. A gate length of 1 μm and width of 20 μm for FET geometry and a power dissipation of 2W are assumed in the table. Depletion-mode devices have extensively employed Buffered FET Logic (BFL) and Schottky Diode FET Logic (SDFL) gate circuit approaches. BFL circuits reported to date have used relatively high pinch-off voltages and level shift diodes and, therefore, have exhibited high power dissipation per gate, as shown in the table. SDFL circuits can achieve high-speed operation comparable to BFL, and, furthermore, result in considerable saving in area/gate and in lower power dissipation. The circuit simplicity of gate design and replacement of FETs with very small Schottky diodes for most logic functions enable depletion-mode devices to achieve high-speed 50-120 ps/gate SSI/MSI circuit levels.

DCL of E-mode devices exhibits both extremely simple circuit design and very low power consumption, and it has been presented as an unrivaled approach for future LSI/VLSI. Making E-mode circuits more complex was probably limited, however, by problems of technology and threshold uniformity. The development of self-aligned E-MESFETs and E-HEMTs has overcome these difficulties, as we show in the following section. LPFL (Nuzillat et al. (28)), which improves logic flexibility of the conventional E-mode device, was proposed to overcome DCL fabrication problems. A range of threshold voltages over two times larger is acceptable by LPFL gates, which can be implemented with so-called "quasi-normally-off" devices. As Table 1 shows, enhancement-mode devices can realize so far the speeds of 100-200 ps/gate with a low power dissipation of 0.02-1 mW/gate and a simple circuit configuration. The 1000 gate/mm^2 scale makes possible a VLSI with 100 kilogates.

Next, we will compare the high-speed performance of various GaAs IC approaches. Table 2 summarizes recently published ring oscillator data. Propagation delays as low as 25 ps/gate are obtained on a submicron gate length ring oscillator (Levy et al. (30)) using E-MESFET DCL. The performance is expected to vary with gate length, gate width, and other factors such as source resistance and parasitic capacitance.

Figure 3 shows the speed to power performance of MESFET and HEMT

logics. For 1.5 μm self-
aligned MESFETs, the minimum
switching delay time is
50 ps and the lowest
power-delay product is
14 fJ. Switching delay is
one-half that of 1 μm
D-MES FET logic at any
power dissipation, and the
power-delay product is
smaller by one order of
magnitude for a switching
delay of 100 ps. Figure 3
also shows the projected
performance of self-
aligned 1 μm gate MESFET
logic. The switching
delay of 17 ps for the
1.7 μm E-HEMT at 77K is
comparable to the top
speed for Josephson
Junction devices of 13 ps
(Gheewala (40)) shown in
this figure. Taking into
account the results of
1.7 μm gate HEMT
technology, the
performance of 1 μm-gate
HEMT logic with a mobility
of 60 000 cm^2/V.s is
projected to a switching
delay below 10 ps with
about 100 μW power
dissipation per stage at
77K.

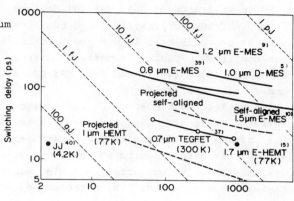

Fig. 3. Speed-power performance of the
self-aligned GaAs MESFET logic and HEMT
logic, compared with currently conventional
GaAs FETs logic and Josephson Junction
logic.

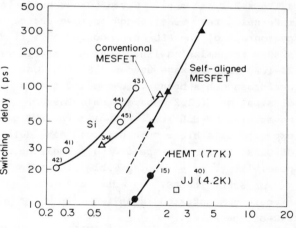

Fig. 4. Switching delay as a function of
gate length.

Figure 4 shows the switching delay as a function of gate length.
Conventional and self-aligned MESFETs at room temperature and HEMTs at
77K are shown here and also compared with Si and Josephson Junction
devices. Under the 1 μm design rule, self-aligned MESFETs are expected
to be below 30 ps at room temperature, and HEMTs to be 10 ps at 77K.
As this figure indicates, self-aligned MESFETs and HEMTs are very
promising candidate for future high-speed VLSI.

2.2 Current status of GaAs MSI/LSI devices

GaAs MSI/LSI complexities have been covered from about 10 to 1000 gates as shown in Fig. 1, and the practicability of IC manufacture has been confirmed. Table 3 summarized current GaAs MSI/LSI device technology (Blum (46)), and GaAs MSI/LSI devices are to be commercially available.

The Hewlett-Packard Company has developed a monolithic MSI GaAs word generator using BFL that operates at data rates from a few bits/s to 5G bits/s for commercial test instrument application (Liechti et al. (47)). This circuit, with 600 active devices, consists of an 8:1 parallel-to-serial converter, a timing generator, control logic, and ECL-interface networks. The circuit generates multiple 8-bit words with dynamic word-length control.

Fujitsu developed GaAs 4-bit arithmetic logic unit (ALU) using a planar ion implantation technique with 2-μm-gate FETs. The basic circuit is a buffered FET logic (BFL) circuit composed of depletion-mode MESFETs and Schottky diodes. Figure 5 shows a microphotograph of the GaAs 4-bit ALU chip, which contains 629 FETs and 225 diodes within a 1.6 x 2.1 mm^2 area. Seven output buffer amplifiers and 76 blocks of internal gates are arranged in rows and columns. Figure 6 shows a photograph of the ALU chip mounted on a flat type package able to drive a 50-Ω transmission line. The package is similar to the conventional 24-lead flat package used for Si ECL ICs. The ALU's switching performance has been evaluated and a data delay of 2.1 ns and power dissipation of 1.2 W have been obtained. This is comparable to that of commercial Si ECL ICs. Reducing the FET gate length to 1 μm, is expected to achieve subnanosecond data delay for ALUs, making the use of GaAs ICs in high-speed data processing applications a realistic possibility.

For SDFL, the LSI circuit demonstrated is an 8x8-bit parallel multiplier developed by Rockwell International. This circuit, consisting of NOR gate full adders and half adders in a regular array, forms the binary product of two 8-bit input words. The planar, localized-implanted fabrication was used for the

Table 3. Present status of GaAs MSI/LSI device technology.

Material :	3" Diameter semi-insulating material
Processing :	Ion implantation
	1 μm lithography
	Dry processing and automated wafer handling
Performance :	Speed, 50–150 ps/gate
	Power, 0.1–1 mW/gate
Process yield :	20–50 % (SSI/MSI)
Complexity :	1000 gate level
Riliability :	Wide temperature operation (-200~+200 °C)
	Radiation hardness (10^5–10^6 rad)

Fig. 5. Microphotograph of the
GaAs 4-bit ALU chip.

Fig. 6. Photograph of the ALU chip
mounted on a flat type package.

2.25×2.17 mm^2 chip. The 8x8-bit parallel multiplier required over
1000 NOR gates (about 3000 FETs and 3000 Schottky diodes) for the
complete circuit. The best performance observed for the 8-bit
multiplier corresponds to a propagation delay of 150 ps/gate at a power
dissipation of about 2 mW/gate. A full 16-bit product would be
available every 5.25 ns.

Rockwell International also developed a high-speed SDFL multimode
divider circuit with 60 gates, capable of frequency division in
divide-by-5, -6, -10, -12, -40, -41, -80, and -82 modes (Walton et al.
(48)). The circuit was implemented using a planar process with
multiple localized ion implantations in semi-insulating GaAs. In the
divide-by-80 and -82 modes, which encompass all other modes of
operation, the circuit operated at a maximum frequency of 1.8 GHz.
Typical power consumption is 150 mW. Functional yield at waferprobe
gave figures indicating as high as 60% of the circuits working at above
1 GHz.

An 8-bit fully decoded RAM test circuit was developed (Bert et al.
(49)) using enhancement-mode GaAs MESFETs with an LPFL circuit.
Correct operation of the circuit has been observed for a supply voltage
varying from 3.5 V to 7 V. An access time of 0.6 ns was measured for a
total power consumption of 85 mW under nominal operating conditions. A

1K-bit static RAM cell array fabricated using self-aligned E/D MESFETs
(Yokoyama et al. (11), (50)) is presented in the next section.

3. ADVANCED TECHNOLOGY FOR HIGH-SPEED GaAs VLSI

This section presents two advanced technologies for achieving DCL
circuit high-speed GaAs VLSI: (1) Self aligned fully-implanted planar
GaAs MESFET technology, and (2) HEMT technology using MBE.

3.1 Self-aligned GaAs MESFET technology

The main problems with DCL technology are the poor reproducibility
of enhancement-mode GaAs MESFETs and their high parasitic resistance.
These make it very difficult to integrate GaAs MESFETs on a very large
scale or to achieve high-speed operation. One approach has been to
develop a novel self-alignment technology for GaAs MESFETs; this would
be analogeous to the polysilicon gate self-alignment technique used in
Si LSI technology. This technology was achieved, for the first time,
by using a high temperature stable TiW silicide Schottky gate metal
system combined with ion implantation technology. There are four major
stages in fabrication, as shown in Fig. 7. (1) A TiW silicide Schottky
gate is formed on an implanted n-type GaAs layer. (2) High dosage
Si^+ implantation is made with the gate acting as an implantation
mask. (3) Annealing is done at $800^\circ C$ for 10 minutes with SiO_2
encapsulation to activate dopants and to form self-aligned n^+
regions. Gate electrical characteristics are not changed during
annealing. (4) Fabrication is completed by ohmic metallization with an
AuGe/Au eutectic alloy. E. Kohn reported (51) that TiW films on GaAs
remain unalloyed up at least $860^\circ C$. Our detailed study showed,
however, that the thermal stability of TiW contacts is not sufficient
after annealing at temperatures exceeding $750^\circ C$ (Yokoyama et al.
(11)).

As packing density has increased, transition metal silicides have
attracted attention for use in Si integrated circuit technology because
of their stability at high temperatures and during device processing
and the ease with which they can be formed and patterned. Here, we
propose to introduce silicide gates in GaAs IC technology. TiW
silicide was formed through rf cosputtering. Backscattering and SIMS
profiles show that TiW silicide does not react with GaAs over a
one-hour period at temperatures up to $850^\circ C$, meaning TiW silicide
annealed on GaAs is much stabler metallurgically than TiW.

Figure 8 shows the TiW silicide Schottky barrier height and ideality factor as functions of annealing temperature. Note that both the barrier height and ideality factor remain comparatively constant after annealing at temperatures of up to 850°C. Moreover, values are very regular.

The ability to etch and generate patterns is one of the most important factors governing the usefulness of metallization in GaAs IC technology. The problem with pattern generation on GaAs wafers is to etch the silicide into fine patterns without etching the GaAs. Figure 9 shows etching rates of TiW silicide and GaAs as a function of the O_2 content in a gas mixture consisting of CF_4 and O_2. Etching

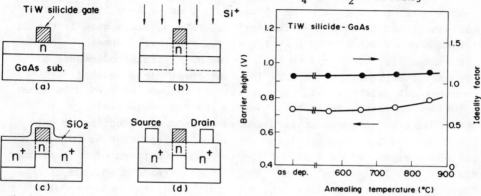

Fig. 7. Fabrication process of the self-aligned GaAs MESFET: (a) Gate metallization, (b) n^+ implantation (c) SiO_2 deposition and annealing, and (d) Ohmic metallization.

Fig. 8. Schottky barrier height and ideality factor of TiW silicide contact on GaAs as a function of annealing temperature.

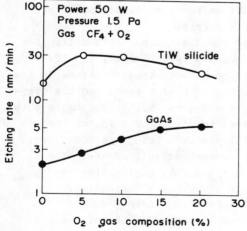

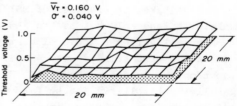

Fig. 10. A map of threshold voltages of the self-aligned E-MESFET.

Fig. 9. Etching rates of TiW silicide and GaAs as a function of the O_2 gas content.

parameters were 50 W for power, 200 V for self-bias voltage, and 1.5 Pa for gas pressure. The TiW silicide etching rate was found to be enhanced by adding a small quantity of O_2 to the CF_4 gas. Adding too much oxygen, however, reduced the etching rate. The etching rate of GaAs increases monotonically with the increase in O_2. Thus, the etching rate of TiW silicide and the ratio of etching selectivity with the GaAs substrate are maximum when O_2 content is 5%.

Selective Si^+ implantation for the n-type layer was made in an HB Cr-doped GaAs substrate at 59 keV with a dosage of 1.10×10^{12} cm^{-2}. The maximum transconductance of self-aligned MESFETs having a gate length of 2 μm and a width of 10 μm, is 1.0 mS (V_T = 0.160 V). The maximum transconductance of conventional FETs having a gate length of 1.2 μm and a width of 50 μm is 1.5 mS (V_T = 0.1 V). When mathematically normalized to a 1 mm wide gate, transconductance of 100 mS for self-aligned MESFETs is about three times that of conventional MESFETs, and source series resistance of 0.75 Ω is about 1/5 times smaller. Reducing parasitic resistance necessarily contributes to enhanced transconductance, as does reduction of gate length due to lateral spread and diffusion of implanted Si to form n^+ regions. Figure 10 is a map of the threshold voltages of self-aligned MESFETs in a 20 x 20 mm^2 area. The average threshold voltage is 0.160 V, with a standard deviation of only 0.040 V. Schottky gate reverse breakdown voltage in self-aligned MESFETs depends on donor density profiles of n^+ regions, because these regions are directly touched to gate electrodes. Therefore, implantation is made with a dosage of 1.7×10^{13} cm^{-2} at as much as 175 keV to maintain a reverse breakdown voltage of at least 6 V and a peak carrier density of 1×10^{18} cm^{-3}.

Let us take, as examples of MSI/LSI devices, a 4x4-bit parallel multiplier and a 1K-bit static RAM cell array.

Figure 11 gives a logic diagram of the 4x4-bit parallel multiplier consisting of 8 full adders, 4 half adders, 16 NOR gates, and 16 input/output buffers. The full adder cell is the basic element in the parallel multiplier. The half adder consists of 8 gates and the output buffer involves two-stage inverters, meaning the multiplier has a gate count of 168. The input signal applied to X's and Y's is inverted by the input buffers and produces partial products using the NOR gates. These products are added by the half adders and full adders. The product output appears at P's. The maximum multiplying time is observed when the product X_4Y_1 appears at P_7. IC fabrication starts with two selective ion implantations in a Cr-doped semi-insulating GaAs substrate. Si was implanted at 59 keV with dosages of 1.1×10^{12} cm^{-2} for E-MESFETs and 2.5×10^{12} cm^{-2}

for D-MESFETs. AuGe/Au ohmic contacts were then formed by the lift-off technique. Schottky gates and ohmic metal layers were also used for first-level interconnects. After the deposition of SiO_2 to provide a metal cross-over structure, a Ti/Au metal layer was formed for second-level interconnects. The completed circuit chip is shown in the microphotograph in Fig. 12. The chip measures 1.5×1.3 mm^2.

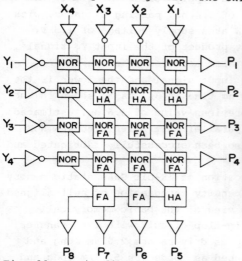

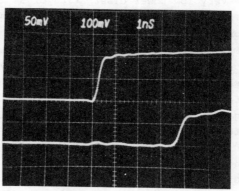

Fig. 11. Logic diagram of 4x4-bit parallel multiplier containing 168 NOR gates.

Fig. 13. Measurement of multiplying time: the upper wave form is the input Y_1 signal, and the lower is the output P_8 signal.

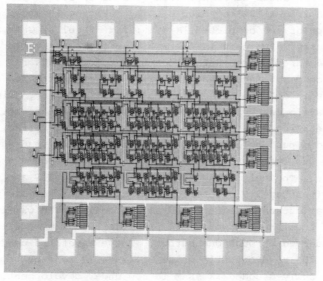

Fig. 12. Microphotograph of 4x4-bit parallel multiplier.

including contact pads, and contains 512 E/D-MESFETs, and the mean
values of threshold voltage were 0.17 V and -0.85 V, respectively. Low
frequency measurements by wafer probe confirmed that the multiplier
functioned correctly. The propagation delay per gate, which depended
on the fixed input code, ranged between 210 ps and 260 ps with a
power dissipation of 0.36 mW/gate and a supply voltage of 1.5 V. We
estimated a 4x4-bit multiplying time of 3.6 ns, passing 14 gates, with
a very low power dissipation of 54 mW at a supply voltage of 1.5 V.
The maximum delay time that the first product of the input Y_1 signal
propagates to output terminal P_8 are also measured as shown in
Fig. 13. The upper waveform is the input signal and the lower is the
output signal. The maximum multiplying time is 3.7 ns as shown in the
photograph, and agrees well at the previous value of 3.6 ns, estimated
from the delay per gate. A 6x6-bit parallel multiplier with 408 NOR
gates in a 2.1 x 1.7 mm^2 area has also been successfully fabricated and
tested.

Next, we will deal with the fabrication of 1K-bit GaAs static memory
cell arrays to demonstrate the high density integration of self-aligned
GaAs MESFETS. Figure 14 shows a top view of the basic memory cell,
which consists of an E/D type FET flip-flop circuit with two transfer
E-FETs. The gates of the E-FETs used as drivers are 2.0 μm long and
10 μm wide, and those of the D-FETs used as loads are 5.5 μm long and
5.0 μm wide. Transfer FETs have gates 2.5 μm long and 4.0 μm wide.
The cell is 50 μm x 39 μm, a size designed for 1K-bit and 4K-bit static
RAMs. Si was implanted at 59 keV with dosages of $1.10 \times 10^{12} cm^{-2}$
for E-FETs and $2.50 \times 10^{12} cm^{-2}$ for D-FETs. Figure 15 shows a
microphotograph of the completed 1K-bit (32 x 32 cells) GaAs static
memory cell array. The chip is 2.1 x 1.7 mm^2. In this array, five
word lines and three paired data lines have bonding pads for access;
others are grounded. Each data line has a column switch and an output
buffer. Fifteen cells in this chip can thus be accessed. Figure 16
shows pulse waveforms typical of the 15 cells to evaluate cell
read/write operation. As shown in the figure, a pull-up FET, two
inverters, and a push-pull circuit were installed for read-out. The
column switch was installed for write-in, which was done directly using
data line pads. The cell array supply voltage was 1.5 V and that of
the output buffer was 3 V. Total cell array current was 110 mA,
indicating that current was supplied to all cells. The waveforms in
Fig. 16 show, from the top, the word pulse (ϕ_x), column pulse (ϕ_y),
and output observed through one buffer. These waveforms confirm that
"0" write-in, "0" read-out, "1" write-in, and "1" read-out operations
were done. A fully-decoded 1K sRAM is now under development.

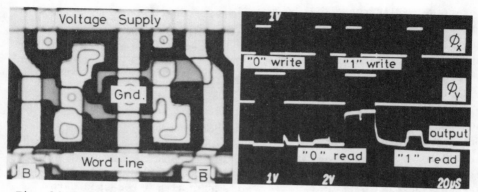

Fig. 14. A top view of the basic memory cell (50 μm x 39 μm).

Fig. 16. Pulse waveforms typical of 15 cells.

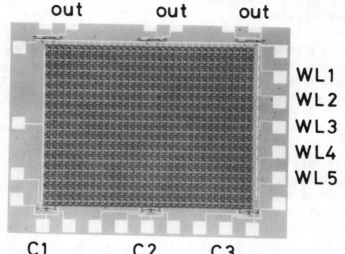

Fig. 15. A microphotograph of the 1K-bit GaAs static memory cell array.

3.2 HEMT technology

HEMT technology has new possibilities for LSI/VLSI with regard to ultra-high-speed and low power dissipation, especially at low temperatures. We have already developed Enhancement (E) and Depletion (D) mode HEMTs, and have successfully integrated E/D type DCL circuits (Mimura et al. (14),(15),(26)). Figure 17 shows a cross section of D-HEMT with a selectively doped GaAs/AlGaAs heterojunction structure. An undoped GaAs and Si-doped n-type AlGaAs layers were successively grown on a semi-insulating GaAs substrate using MBE. Because of the higher electron affinity of GaAs, free electrons in the AlGaAs layer are transferred to the undoped GaAs layer, where they form a two-dimensional high mobility electron gas (2DEG) within 100 Å of the interface. Electron mobility and sheet electron concentration (N_s) in the heterostructure are shown as a function of temperature in Fig. 18. (Hiyamizu et al. (52)). As temperature decreased, electron mobility which was 8070 cm^2/V.s at 300K, increased dramatically and reached 121000 cm^2/V.s at 77K due to reduced phonon scattering. A further increase with a considerable gradient occurred even below 50K, and a maximum value of 260000 cm^2/V.s was attained at 5K. Sheet electron concentration decreased with decreasing temperature until it became constant below 150K. The almost constant value of about 4.9×10^{11} cm^{-2} below 150K corresponds to that of 2DEG at the interface, since this value agrees well with the value of N_s determined by Shubnikov-de Haas measurement at 4.2K. Apparent excess carriers above

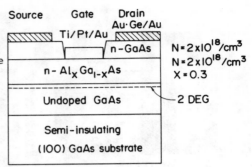

Fig. 17. Cross section of depletion mode HEMT with a recessed gate.

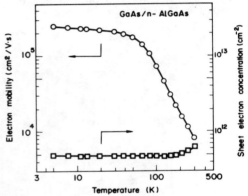

Fig. 18. Electron mobility and sheet electron concentration in GaAs/n-AlGaAs as a function of a temperature.

150K are attributed to free electrons which are thermally excited from relatively deep donors (70 meV) in n-type $Al_{0.3}Ga_{0.7}As$. Figure 19 shows annual 2DEG mobilities reported so far for selectively doped heterostructures and modulation-doped superlattices starting with 1978, when modulation-doping in superlattices was first demonstrated. Open circles indicate mobilities in selectively doped heterostructures (SH) at 77K and solid circles show mobilities at 4.2 - 10K. During 1978 and 1979, mobility remained rather low. It began increasing rapidly, however, when the first HEMT was developed. Mobility apparently continues to increase, and has achieved values of 1.2 million cm^2/V.s at 5K, and 2 million cm^2/V.s at 5K under photo excitation (Hiyamizu et al. (53)).

Figure 20 shows the energy band diagram of D- and E-HEMTs in thermal equilibrium. The n-type AlGaAs layer of D-HEMT is completely depleted by two mechanisms: Surface depletion results from the trapping of free electrons by surface states, and interface depletion results from the

transfer of electrons to undoped
GaAs. The Fermi level of the gate
metal is matched to the pinning
point, which is 1.2 eV below the
conduction band. With the reduced
AlGaAs layer thickness, electrons
supplied by donors in the AlGaAs
layer are insufficient to pin the
surface Fermi level. Therefore,
the space charge region extends
into the   undoped GaAs layer and,
as a result, band bending results
in an upward direction and the
two-dimensional electron gas does
not appear, as shown in the
figure. When a positive
voltage V_{GS} higher than the
threshold voltage is applied to

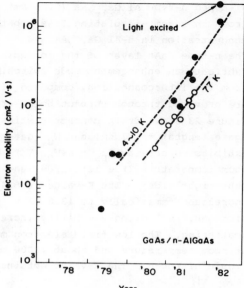

Fig. 19. 2DEG mobility are improved
in selectively doped GaAs/n-AlGaAs
as a function of year.

the gate, electrons accumulate at
the interface and form a
two-dimensional electron gas,
shown by the broken lines, producing an E-HEMT. In Fig. 21, transfer
characteristics of the square root of the drain saturation current
versus gate voltage, $\sqrt{I_{DS}}$ versus V_{GS}, are plotted for both E and
D-HEMTs. Both devices exhibit square law $I_{DS} = K(V_{GS} - V_T)^2$
characteristics. The K-value is given by $\varepsilon \mu_n W_G / 2aL_G$ in the low
field region. Here, ε is the dielectric constant of the
Al_xGa_{1-x} As layer, μ_n is the field effect mobility, and a is the
thickness of the Al_xGa_{1-x} As layer. By lowering the temperature to
77K, a dramatic K-value increase by a factor of 3 is observed. This
increase is due to increased electron mobility at low temperatures.
Low field electron mobility was found from Hall measurements to be
around 6000 cm^2/V.s at 300K and around 20000 cm^2/V.s at 77K. Note
that the threefold improvement in K-value at 77K is about the same as
that of low field electron mobility at 77K, although the K-value is
measured in the high field region (average field within the channel:
4 kV/cm) where velocity saturation effects become very significant. At
$V_{DS} = 1.5$ V and $V_{GS} = 0.7$ V, the E-HEMT exhibits g_m of 193 mS/mm at
300 K; g_m increases to 409 mS/mm at 77K. This g_m value at 77K is
the highest ever reported for any field effect devices.

Figure 22 shows a cross section of an experimental inverter
structure with enhancement-mode switching and depletion-mode load
HEMTs. The epilayer, consisting of undoped GaAs (0.8 μm-thick),

Si-doped n-type $Al_xGa_{1-x}As$ (0.06 μm) and n-type GaAs (0.05 μm), is grown on a semi-insulating GaAs substrate by MBE. The doping concentration in $n-Al_xGa_{1-x}As$ and n-type GaAs is 2 x 10^{18} cm^{-3}. The n-type GaAs layer of the top epilayer is removed by etching to fabricate an enhancement-mode switching device. Electrical connections from the interconnection, composed of Ti/Pt/Au, to the device terminal are provided through contact holes etched in the SiO_2 film.

Figure 23 shows drain characteristics of the switching HEMT with L_G (gate length) = 1.7 μm, and W_G (gate width) = 33 μm. The device exhibits square-law $I_{DS} = K(V_{GS}-V_T)^2$ characteristic at both room temperature (the left side) and the temperature of liquid nitrogen (the right side). The K-value is 5.4 mA/V^2 at room temperature and increases dramatically to 13.6 mA/V^2 at the temperature of liquid nitrogen, reflecting a mobility increase due to reduced phonon scattering. The low field electron mobility was about 6000 $cm^2/V.s$ at room temperature and about 20000 $cm^2/V.s$ at the temperature of liquid nitrogen. The K-value obtained with HEMT is 2.2 times (at room

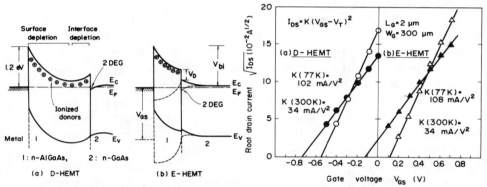

Fig. 20. Energy band diagram of the
(a) D- and (b) E-HEMTs.

Fig. 21. Transfer characteristics of the square root of the drain saturation current versus gate voltage of the D- and E-HEMTs.

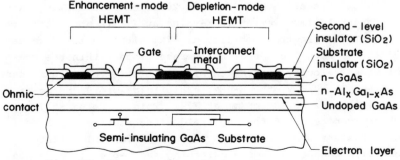

Fig. 22. Cross sectional view of HEMT inverter.

temperature) and 5.8 times (at the temperature of liquid nitrogen) higher than that for a 1-μm-gate GaAs MESFET (K=0.7 mA/V^2 for W$_G$=10 μm) (Eden et al. (23)), in spite of the HEMT's greater gate length.

We next studied the optimized conditions for selective dry etching of GaAs to AlGaAs by using MBE-grown GaAs-AlGaAs heterojunction materials, where the etching satisfied the following conditions: 1) the etching rate of AlGaAs is very low, 2) the surface morphology of GaAs and AlGaAs is clean and smooth, and 3) the resist-mask is sufficiently durable for dry etching. Figure 24 shows typical etching depth versus etching time characteristics in GaAs (60 nm thick)-AlGaAs heterojunction materials at a gas composition ratio of P$_{CCl_2F_2}$/P$_{He}$=1 (Hikosaka et al. (54) (55)). The properties of GaAs material under the same conditions are included for comparison. The most interesting result in the figure is that an initial time delay of about 10 seconds on the etching time is obvious in GaAs. This generally implies that the delay is caused by ion sputtering of native oxide on the material. In Fig. 24, presputtering at a power density of 0.35 W/cm^2 was initially done for 8-9 seconds for etching at 0.18 W/cm^2, because the unreacted overlayer on the GaAs surface cannot be removed by ion sputtering at this power

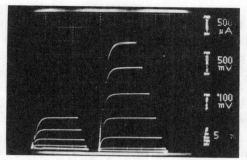

Fig. 23. Drain current-voltage characteristics.

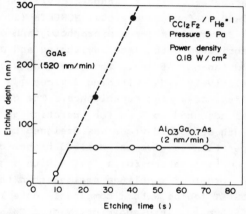

Fig. 24. Etching depth as a function of etching time.

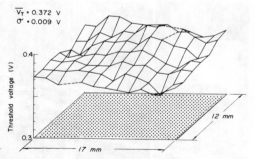

Fig. 25. A map of threshold voltages of E-HEMT.

density. A significantly high selectivity ratio of more than 200 was
achieved at a power density of 0.18 W/cm^2, where the etching rate of
AlGaAs was as low as 2 nm/min and that of GaAs was about 520 nm/min.
An increased power density up to 0.35 W/cm^2 also remarkably increases
the etching rate of AlGaAs (30 nm/min) compared to the rate at
0.18 W/cm^2, though the rate of GaAs increased. As a result of the
low etching rate of AlGaAs material at a power density of 0.18 W/cm^2,
the desired selective dry etching is achieved. Figure 25 shows a map
of the threshold voltage of an E-HEMT with a gate length of 2 μm, a
width of 50 μm, in an area of 17 x 12 mm^2 at room temperature, The
average threshold voltage is 0.372 V with a standard deviation of only
0.009 V, 2.5% of the average threshold voltage. The present selective
dry etching promises excellent uniformity in controlling device
parameters, with a standard deviation about 1/3 smaller than that of
50-80 mV for conventional MESFETs (Zucca et al. (56)).

Figure 26 shows a microphotograph of a 27-stage ring oscillator
fabricated with E-mode switching and D-mode load HEMTs, with a fan-in
and fan-out of one. The gate length of the switching HEMT is 1.7 μm.
Relatively wide switching devices (33 μm) are used to minimize the
effect of wiring capacitance. The gate width of D-mode load devices
was adjusted to 7 μm for operation at room temperature, and the gate
width of load devices was designed to 13 μm for low temperature
operation to give the desired higher current. Power dissipation at
room temperature for a device with a gate length of 1.7 μm is 0.46 mW,
one order of magnitude smaller than that of currently achievable GaAs
MESFET logics with a comparable gate length at the same switching time
(56.5 ps). This superiority over GaAs MESFET logics in speed-power
performance of the device at room temperature device partly results
from the higher electron mobility (6000 cm^2/V.s) of HEMTs compared to

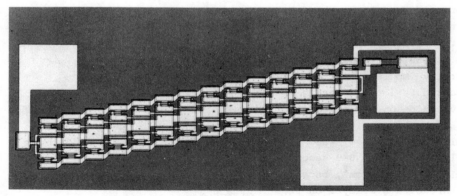

Fig. 26. Fabricated 27-stage HEMT ring oscillator
with an output buffer and a probe resistor.

MESFETs (4500 cm^2/V.s). At the temperature of liquid nitrogen, a
switching time of 17.1 ps with a power dissipation of 0.96 mW has been
achieved. The power dissipation of 0.96 mW at the temperature at
liquid nitrogen is twice as much due to the higher current in the wider
gate of the D-mode load device. The switching time of 17.1 ps is the
lowest ever reported in Si and GaAs logic technologies and is
comparable with that of Josephson Junction logic (13 ps). More
recently, a switching time of 16.7 ps per stage at room temperature and
12.8 ps per stage at the temperature of liquid nitrogen have been
achieved in a 27-stage ring oscillator with an E-HEMT with a 1.1 μm
gate. Taking into account the results of 1.7 μm-gate HEMT technology,
1 μm-gate HEMT logic with an improved mobility of about 60000 cm^2/V.s
can be expected to achieve switching delay of 10 ps with about 100 μW
power dissipation per stage at the temperature of liquid nitrogen, as
shown in Fig. 3.

4. FUTURE PROSPECTS OF GaAs VLSI

The field of high-speed GaAs IC is expanding rapidly. Most GaAs IC
applications lie in future high-performance digital and analog systems,
such as computers and communication systems. These systems require
high-speed logic and memory covering a wide range of complexity from
SSI to VLSI. Future high-speed digital electronic systems include
fiber-optical communication networks, radar signal processors,
prescalers for counters, frequency synthesizers, and digital phase lock
loops. Test instruments that generate, receive, and analyze data at
rates greater than 1G bits/s are also needed. These needs can be met
with GaAs monolothic ICs operating at clock frequencies beyond the
reach of present Si ICs.

Recent large-scale general-purpose computers, using Si-based
technology, have already achieved speeds of 30 MIPS (Million
Instructions Per Second). Figures 27 and 28 show future requirements
of large-scale general-purpose computer (Abe et al. (57)) and super
computers (Misugi et al. (58)). Improvement of computer performance is
under way and computer speed will reach 100 MIPS for general-purpose
computers and 1000 MFLOPS (Million FLoating-point Operations Per
Second) for supercomputers at the end of the 1980s. The requirements
for higher speed processing is expanding even faster than actual
computer improvement, however, especially in scientific fields such as
nuclear physics, weather forecasting, and resources exploration. It
may be difficult for Si-based technology to achieve the above

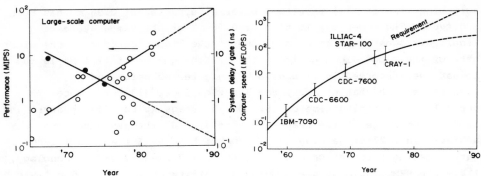

Fig. 27. Evolution of large scale general purpose computers.

Fig. 28. Evolution of supercomputer

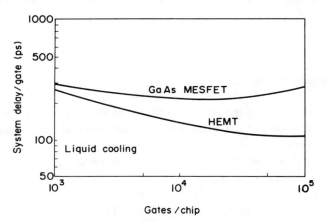

Fig. 29. Projected system delay calculated as a function of the number of gates.

requirements. GaAs LSI/VLSI, however, seem the most promising candidates to satisfy CPU and cache memory requirements for the mainframes of future computers. Today's computer systems are constructed by stacking printed circuit boards. System delays result mainly from two sources: chip delay and external wiring delay. Figure 29 shows the system delay per gate which is the sum of the chip delay and the external wiring delay. Although for small-scale integration, most delays result from external wiring, chip delay premodinates for large-scale integration. Optimum system performances are system delays of 200 ps at 10 kilogates for GaAs MESFET VLSI at room temperature, and 100 ps at 100 kilogates for HEMT VLSI at the liquid nitrogen temperature. These system performances are expected to reach speeds higher than 100 MIPS for future large-scale computer requirements.

Projected GaAs IC complexity is shown by the broken lines in Fig. 1. As this figure indicates, future high-speed GaAs VLSI complexity can be realized at an approximate threefold annual increase. GaAs MESFET VLSI complexity will meet the evolution line of a threefold increase at about 10 kilogates, and HEMT VLSI at 100 kilogates. The complexities of 10 and 100 kilogates just agree at complexities to realize optimized system performances of MESFET and HEMT VLSIs, respectively. We believe that the advanced device technologies of self-aligned MESFETs and HEMTs presented here will contribute greatly to GaAs VLSI evolution.

5. CONCLUSIONS

GaAs SSI/MSI have already been put into practical use for high-speed digital electronic systems, based on BFL and SDFL technologies. DCL technology is suitable for high-speed GaAs VLSI for future large-scale computer applications. One of the most important breakthroughs for DCL is self-aligned fully-implanted GaAs MESFETs technology. This was achieved, for the first time, by using TiW silicide gates stable at 850°C, combined with ion implantation technology. By using this technology, we first demonstrated 4x4-bit and 6x6-bit parallel multipliers and a 1K-bit static RAM cell array.

HEMTs are very promising devices for VLSI, especially operating at liquid nitrogen temperature, because of their ultra-high-speed and low power dissipation. E/D type Direct Coupled HEMT Logic achieved a switching time of 17.1 ps at 77K, despite a comparatively wide 1.7 μm gate length, and precisely controllable technology for fabricating HEMT devices has also been developed. Selective dry etching of GaAs to AlGaAs can be achieved with a ratio exceeding 200 by using an etching gas composed of CCl_2F_2 and helium. The projected HEMT performance target suitable for VLSI is a switching delay below 10 ps with a power dissipation of about 100 μW per stage under 1 μm design rule technology.

Using the experimental data on self-aligned MESFET and HEMT logic, we projected optimized system performance of 200 ps at 10 kilogates integration with GaAs MESFET VLSI at room temperature, and 100 ps at 100 kilogates with HEMT VLSI at liquid nitrogen temperature. These system performances will achieve speeds higher than 100 MIPS for future large-scale computer requirements.

ACKNOWLEDGEMENTS

We would like to thank Dr. T. Misugi, Dr. O. Ryuzan, H. Ishikawa, T. Kotani, M. Mukai, and Dr. M. Fukuta for their encouragement and support. We would also like to express our gratitude to Dr. S. Hiyamizu and A. Shibatomi for their many useful discussions.

This work was partially supported by the Ministry of International Trade and Industry of Japan.

REFERENCES

(1)   R. L. Van Tuyl and C. A. Liechti, IEEE J. Solid-State Circuits SC-9 (1974) 269.

(2)   G. Nuzillat, presented at the 7th European Solid State Circuit Conference (FSSCIRC-81) (1981) 65.

(3)   R. L. Van Tuyl, C. A. Liechti, R. E. Lee and E. Gowen, IEEE J. Solid-State Circuits SC-12 (1977) 485.

(4)   K. Suyama, H. Kusakawa, S. Okamura, S. Yamamura, and M. Fukuta, IEEE GaAs IC Symp., (1981) Paper 4.

(5)   R. C. Eden, B. M. Welch, and R. Zucca, ISSCC Dig. Tech. Papers (1978) 68.

(6)   S. I. Long, F. S. Lee, R. Zucca, B. M. Welch, and R. C. Eden, IEEE Trans. Microwave Theory and Tech., MTT-28 (1980) 466.

(7)   F. S. Lee, E. Shen, G. R. Kaelin, B. M. Welch, R. C. Eden and S. I. Long, IEEE GaAs IC Symp., (1980) Paper 3.

(8)   F. S. Lee, R. C. Eden, S. I. Long, B. M. Welch, and R. Zucca, in Proc. IEEE Int. Conf. Circuits and Computers, (1980) 697.

(9)   H. Ishikawa, H. Kusakawa, K. Suyama, and M. Fukuta, ISSCC Dig. Tech. Papers (1977) 200.

(10) N. Yokoyama, T. Mimura, M. Fukuta and H. Ishikawa, ISSCC Dig. Tech. Papers (1981) 218.

(11) N. Yokoyama, T. Ohnishi, K. Odani, H. Onodera, and M. Abe, IEDM Dig. Tech. Papers (1981) 80.

(12) K. Suyama, H. Shimizu, S. Yokogawa, Y. Nakayama, and A. Shibatomi, The 14th Conf. (1982 International) on Solid State Devices Dig. Tech. Papers (1982) 185.

(13) Y. Nakayama, K. Suyama, H. Shimizu, S. Yokogawa, and A. Shibatomi, to be presented at IEEE GaAs IC Symp. (1982)

(14) T. Mimura, S. Hiyamizu, T. Fujii, and K. Nanbu, Japan, J. Appl. Phys., 19 (1980) L225.

(15) T. Mimura, K. Joshin, S. Hiyamizu, K. Hikosaka, and M. Abe, Japan. J. Appl. Phys., 20 (1981) L598.

(16) M. Abe, T. Mimura, N. Yokoyama, and H. Ishikawa, IEEE Trans. Microwave Theory and Tech. MTT-30 (1982) 992.

(17) P. N. Tung, D. Delagebeaudeuf, M. Laviron, P. Delescluse, J. Chaplart, N. T. Linh, Electronics Lett. 18 (1982) 109.

(18) R. C. Eden, Proc. IEEE, 70 (1982) 5.

(19) P. M. Solomon, Proc. IEEE, 70 (1982) 489.

(20) H. Kroemer, Proc. IEEE, 70 (1982) 13.

(21) M. Heiblum, Solid-State Electron. 24 (1981) 342.

(22) N. Yokoyama, T. Mimura, and M. Fukuta, IEEE Trans. Electron Devices ED-27 (1980) 1124.

(23) R. C. Eden, B. M. Welch , R. Zucca, and S. L. Long, IEEE Trans. Electron Devices ED-26 (1979) 299.

(24) R. Zuleeg, J. K. Notthoff, and K. Lehovec, IEEE Trans. Electron Devices, ED-25 (1978) 628.

(25) C. O. Bozler and G. D. Alley, Proc. IEEE, 70 (1982) 46.

(26) T. Mimura, S. Hiyamizu, K. Joshin, and K. Hikosaka, Japan. J. Appl. Phys., 20 (1981) L317.

(27) R. Dingle, H. L. Störmer, A. C. Gossard, and W. Wiegmann, Appl. phys. Lett., 33 (1978) 655.

(28) G. Nuzillat, E. H. Perea, G. Bert, F. D. Kavala, M. Gloanec, M. Peltier, T. P. Ngu, and C. Arnodo, IEEE J. Solid State Circuits, SC-17 (1982) 569.

(29) R. E. Lundgren, C. F. Krumm, and R. L. Pierson, presented at 37th Annu. Dev. Research Conf., Boulder, Co., (1979).

(30) H. M. Levy, R. E. Lee, and R. Saldler, 40th Annu. Dev. Research Conf., (1982) IV B-2.

(31) R. L. Van Tuyl and C. Liechti, IEEE ISSCC, Dig. Tech. Papers (1979) 20.

(32) B. M. Welch, Y. D. Shen, R. Zucca, R. C. Eden, and S. I. Long, IEEE Trans, Electroh Devices, ED-27 (1980) 1116.

(33) K. Suyama, H. Kusakawa and M. Fukuta, IEEE Trans. Electron Devices, ED-27 (1980) 1092.

(34) T. Mizutani, N. Kato, S. Ishida, K. Osafune, and M. Ohmori, Electron. Lett., 16 (1980) 315.

(35) G. Nuzillat, F. Damay-Kavala, G. Bert, and C. Arnodo, Proc. Inst. Elec. Eng., 127 (1980) 287.

(36) G. Nuzillat, G. Bert, T. P. Ngu and M. Gloance, IEEE Trans. Electron Devices, ED-27 (1980) 1102.

(37) P. N. Tung, P. Delescluse, D. Delagebeaudeuf, M. Laviron, J. Chaplart, and N. T. Linh, Electronics Lett. 18 (1982) 517.

(38)   K. Lehovec and R. Zuleeg, IEEE Trans. Electron Devices, ED-27
       (1980) 1074.
(39)   T. Mizutani, M. Ida and M. Ohmori, Digest of IEEE/MTT-S 1st
       Speciality Conf, on Gigabit Logic for Microwave Systems (1979) 93.
(40)   T. R. Gheewala, IEEE J. Solid-State Circuits, SC-14 (1979) 787.
(41)   M. P. Lepselter IEDM Dig. Tech. Papers (1980) 42.
(42)   M. P. Lepselter, IEEE GaAs IC Symp. (1981) Paper 5.
(43)   K. Ohwada, Y. Omura and E. Sans, IEDM Dig. Tech. Papers (1980)
       756.
(44)   K. Nishiuchi, H. Shibayama, T. Nakamura, T. Hisatsugu,
       H. Ishikawa and Y. Fukukawa, ISSCC Dig. Tech. Papers (1980) 60.
(45)   I. Ito, H. Ishikawa, and Y. Fukukawa, Proc. of 12th Conf. on Solid
       State Devices (1980) 9.
(46)   F. A. Blum, IEEE GaAs IC Symp. (1981) Paper 6.
(47)   C. A. Liechti, G. L. Baldwin, E. Gowen, R. Joly, M. Namjoo, and
       A. F. Podell, IEEE Trans. Microwave Theory and Tech., MTT-30
       (1982) 998.
(48)   E. R. Walton, Jr., E. K. Shen, F. S. Lee, R. Zucca, Y.-D. Shen,
       B. M. Welch, and R. Dikshit, IEEE Trans. Microwave Theory and
       Tech., MTT-30 (1982) 1020.
(49)   G. Bert, J.-P. Morin, G. Nuzillat, and C. Arnordo, IEEE Trans.
       Microwave Theory and Tech. MTT-30 (1982) 1014.
(50)   N. Yokoyama, T. Ohnishi, K. Odani, H. Onodera, and M. Abe, IEEE
       Trans. Electron Devices, to be published, (Oct. 1982)
(51)   E. Kohn, 1979 IEDM, Tech. Dig. (1979) 469.
(52)   S. Hiyamizu, T. Mimura, and T. Ishikawa, Proc. 13th Conf. on
       Solid State Devices (1981); Japan. J. Appl. Phys. 21, Supplement
       21-1 (1982) 161.
(53)   S. Hiyamizu, 2nd Int. Symp. on Molecular Beam Epitaxy and Related
       Clean Surface Techniques (1982) Paper A-7-1.
(54)   K. Hikosaka, T. Mumura and K. Joshin, Japan. J. Appl. Phys., 20
       (1981) L847.
(55)   K. Hikosaka, T. Mimura, K. Joshin, and M. Abe, The Electro
       chemical Society, Inc. (1982) No. 163.
(56)   R. Zucca, B. M. Welch, C.-P. Lee, R. C. Eden, and S. I. Long,
       IEEE Trans. Electron Devices, ED-27 (1980) 2292.
(57)   M. Abe, T. Mimura, N. Yokoyama, and H. Ishikawa, IEEE GaAs IC
       Symp. (1981) Paper 7.
(58)   T. Misugi and K. Kurokawa, 1st Annual Phoenix Conf. on Computers
       and Communications (1982) 419.

Metal Organic Vapour Phase Epitaxy: The Key Issues

Jean HALLAIS

Laboratoires d'Electronique et de Physique Appliquée
3, avenue Descartes, 94450 Limeil-Brévannes, France

SUMMARY

Metal organic vapour phase epitaxy is becoming more and more used in research on compound semiconductors (III-V and II-VI's) for device-oriented growth.

This paper reviews the development of the technique for the growth of $Al_xGa_{1-x}As$ - GaAs and $Ga_{1-x}In_xAs_{1-y}P_y$ - InP heterostructures using atmospheric or reduced pressure operating conditions. The problems of purity and carbon contamination are discussed for GaAs and $Al_xGa_{1-x}As$ with emphasis on oxygen contamination for the aluminium containing layers. Recent results on lasers are reported, especially the low threshold current density GaAs - GaAlAs quantum well lasers and the quaternary ones. Feasability of quantum wells with concentration changes at the interface over less than one unit cell is shown. Finally, some problems which remain to be solved are discussed and one attempt is made to anticipate the future development of MO-VPE.

1 INTRODUCTION

Epitaxial growth of compound semiconductors layers has been studied for about thirty years and it has become a major step in the device fabrication technology.

Several techniques, namely the liquid phase epitaxy (LPE), the vapour phase epitaxy based on chloride transport (VPE), more recently the molecular beam epitaxy (MBE) and the metalorganic vapour phase epitaxy (MO-VPE) have been successfully developed. This large variety of techniques gives also evidence for the difficulty of achieving epitaxial structures of increasing complexity.

A good approximation of the relative importance of the research effort on each technique can be deduced from a survey of the open literature. The proceedings of the conferences on Gallium Arsenide and Related Compounds (published by the Institute of Physics, London) have been taken as a reference. Figure 1 shows the percentage of papers dealing with each growth technique over the last ten years period.

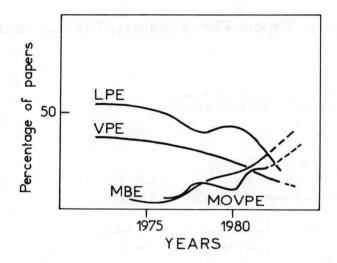

FIGURE 1 : Percentage of contributions dealing with each growth technique over the last ten years.

General trends are :
(i) a decrease of the older activities on LPE and VPE which benefits to the the more recent MBE and MO-VPE studies,
(ii) LPE is no more the unique method for growing (Al,Ga)Ga - GaAs hetero-structures but is challenged by MBE and MO-VPE,
(iii) LPE and VPE have reached maturity and are now used in industrial opera-tions,
iv) there is some competition between MBE and MO-VPE.

In fact, the actual requirements for epitaxial layers are no more only control of thickness uniformity, purity and doping, but also the possibility of growing ternary or quaternary multiple heterostructures with abrupt inter-faces and ultra-thin layers of about 100 Å or less. As it will be shown, MO-VPE seems capable of fulfilling these requirements. The advance is rather fast and table I lists some milestones in the history of MO-VPE. High performance optoelectronic and high speed devices based on $Al_{1-x}Ga_xAs$ - GaAs and $Ga_{1-x}In_xAs_{1-y}P_y$ -InP have already been obtained, some work on $Cd_{1-x}Hg_xTe$ - CdTe is in progress.

State of the art results will illustrate the versatility of the technique but also point out the problems which remain to be solved.

1968 - 1971	Early work by H. Manasevit
	III-V and II-VI compound semiconductors on insulators
1972	ICGC Marseille
	The MANASEVIT's growth technique is acknowledged by the crystal growth community (1)
1975	High purity GaAs demonstrated by Y. Seki (2)
	First report by S.J. BASS of device quality GaAs for reflexion mode photocathodes and MESFET's (3)
1976	(Al,Ga)As - GaAs heterostructures for transmission mode photocathodes by J.P. ANDRE et al. (4)
1977	. D.H. lasers by R.D. DUPUIS and P.D. DAPKUS (5)
	. Intense activity in several laboratories
	. First report on GaAs growth by low pressure MO VPE by J.P. DUCHEMIN (6)
1978	. Quantum well (Al,Ga)As - GaAs heterostructure lasers by R.D. DUPUIS, P.D. DAPKUS and N. HOLONYAK (7)
	. InP growth by LP MO VPE by J.P. DUCHEMIN (8)
1980	. Quaternary laser by J.P. HIRTZ (9)
1981	First International Conference on Metal organic Vapour Phase Epitaxy.

<u>TABLE I : MILESTONES IN MO-VPE</u>

2 CHEMISTRY OF THE GROWTH AND REACTOR DESIGN

The basis of the process to grow a compound AB is the simultaneous thermal decomposition of volatile compounds of A and B. Namely, the metal-organic compounds are the alkyls for the groups III and II elements and the hydrides are the source for the groups V and VI elements.

These types of sources work well for GaAs, (Al,Ga)As and GaP. In the case of GaAs growth, using trimethylgallium (TMGa) carried by hydrogen and arsine, it has been shown (10) that the decomposition of TMGa is complete at 700°C according to the reaction :

$$2 \ Ga \ (CH_3)_3 \ + \ 3 \ H_2 \ \longrightarrow \ 2 \ Ga \ + \ 6 \ CH_4 \ . \hspace{2cm} (1)$$

The degree of decomposition of AsH_3 at the same temperature following :

$$4 \ AsH_3 \ \longrightarrow \ As_4 \ + \ 6 \ H_2 \hspace{3cm} (2)$$

is close to 60 %. To allow the epitaxial growth of GaAs, one generally uses a large arsine over TMGa mole fractions ratio. No surface kinetic limitation of the growth is observed, even at reduced pressure, for temperatures ranging from 600 to 800°C (**11**). The growth is then only governed by mass transfer and the gas flow dynamics is important for the optimum reactor design.

Clearly, an unambiguous model to describe the growth mechanism is not available yet ; one can simply rule out the previous model which assumed an adsorption of the TMGa and arsine molecules on surface sites and reactions between these molecules (**12**). An attractive hypothesis is that the TMGa molecules can simply be seen as a source of gallium atoms which diffuse through the boundary layer towards an arsenic rich surface. The growth mechanism in MO-VPE would then be comparable to the MBE process (**13**).

The design of the growth chamber is rather simple, based either on horizontal or vertical (Figure **2**) geometry. R.F. heating is mostly used whereas the application of radiations or Joule heaters has also been reported.

Using a mixture of TMGa and TMAl (trimethylaluminium) instead of TMGa allows the growth of $Al_xGa_{1-x}As$ without other major changes

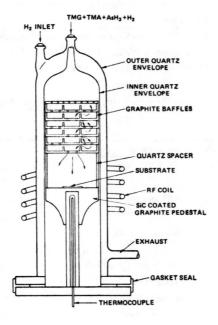

FIGURE 2 :

Schematic diagram of vertical OM-VPE reactor with graphite baffles to enhance O_2 and H_2O with TMAl to reduce oxygen contamination (after 38).

The growth of In containing materials is troubled by a detrimental elimination reaction between trimethyl (or ethyl) indium and the hydrides (AsH_3

or PH_3). When mixing triethylindium (TEIn) with AsH_3, this type of Lewis acid-base reaction leads to a less volatile liquid whereas mixing TEIn with PH_3 gives a solid (14,15). Early solutions to this problem involved separation of group III alkyls from AsH_3 until they approach the hot zone for $Ga_{1-x}In_xAs$ growth. Unfortunately, the elimination of In is never completely avoided and it remains difficult to control the gas phase composition in the reaction chamber (16). Later solutions involved either the replacement of AsH_3 or PH_3 by respectively trimethylarsine (TM As) [17] or trimethyl-phosphine (TMP) (18) or growth at low pressure (8). Reduced pressure growth requires a slightly more complicated equipment than growth at atmospheric pressure, (figure 3), and the use of a pre-pyrolysis of the phosphine before mixing was found necessary. This technique is successfully applied to the growth of InP and $Ga_{1-x}In_x As_{1-y}P_y$ for devices (19).

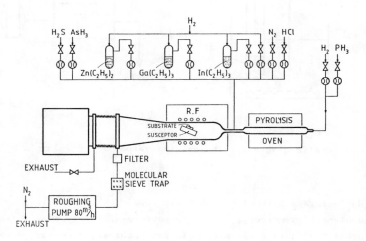

<u>FIGURE 3</u> : LP - MO-VPE reactor for the growth of InP and related
compounds (after 19).

The use of adducts formed of group III and group V alkyls looks like an interesting alternative. The growth of GaAs from monochlorodialkyl gallium added to trialkyl arsine (20) or of InP from $(CH_3)_3 In-P (CH_3)_3$ (18) has been demonstrated. Purity or stoichiometry problems were reported but these difficulties seem to have been recently overcome using the adduct as a metal source and an excess of hydride for the group V element (15,21). For instance, this leads to good quality InP prepared at 630°C by mixing a trimethyl indium triethylphosphine carried by nitrogen and PH_3 carried by a large excess of hydrogen. The reactor design is quite simple (figure 4) and does not require any more equipment than the usual MO VPE atmospheric pressure set-up (15).

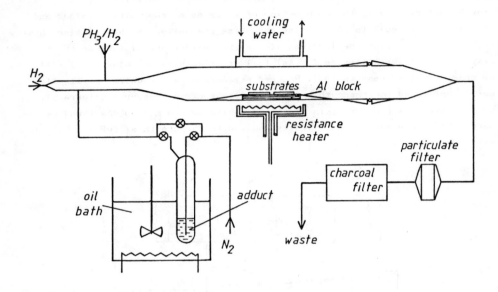

FIGURE 4 : Schematic diagram of the growth system (after 15)

$Cd_{1-x}Hg_xTe$ layers have also been grown from diethyltellurium, dimethyl-cadmium and elementary mercury. Growth is performed at the lowest possible temperature (about 410°C) to avoid interdiffusion problems and seems to be limited by the decomposition of the tellurium compound (22).

Many other compound semiconductors have been grown by MO-VPE (23) but the work is less advanced than for the three major systems considered above.

3 PURITY AND DOPING

GaAs and $Al_xGa_{1-x}As$ layers have been far more investigated in terms of purity and doping than any other semiconductor compound. One has nevertheless to keep in mind that high purity InP and (Ga,In)-(As,P) quaternaries have also been obtained (19).

Doping is not really a problem and many dopants are well controlled. N-type dopants like Si, S, Se are suitable for GaAs, Si and Se are convenient for $Al_{1-x}Ga_xAs$. Zinc is commonly used to obtain P-type GaAs or (Al,Ga)As. The behaviour of these impurities versus growth temperature and reactant mole fractions has already been reviewed (11) and only slow diffusing impurities for P-type doping like Mg (24) or possibly Be (25) are less known.

A lot of work has been devoted to purity which raises the following questions :

(i) which impurities are involved in the poor batch to batch reproducibility for commercially available metalorganics and hydrides ?

(ii) is carbon a severe contamination source ?

(iii) can one eliminate any oxygen and water contamination when growing $Al_{1-x}Ga_xAs$?

It is well known that the background doping can change from P to N-type with increasing temperature and/or with increasing arsenic to gallium ratio (11). This has tentatively been related to impurities carried by the TMGa (or TEGa) and the arsine, assuming Si (2), Zn (26), Mg (27) and C (2). This attribution mostly relies on photoluminescence spectra as shown in figure 5 (38) In that example, first consider the spectrum recorded at 4 K for sample 1 (full line), which is typical for layers grown at low temperatures ($<$ 680°C) and having about 7.10^{13} cm^{-3} dopant concentration, n type. A sharp excitonic structure is observed which has been also reported by other authors (2). It is due to the recombination of free exciton (F.E.), of excitions bound to neutral donors (D_o, X), or free holes with an electron bound to a donor (D_o, h), and of an exciton bound to a neutral acceptor (A_o, X) (29). The lines observed in the 830 nm region at 1.489 and 1.486 eV correspond to band acceptor (BA) and donor acceptor (DA) recombination, the acceptor involved being zinc. This interpretation is based on the fact that with increasing temperature or excitation density the DA band disappears.

The other spectrum in figure 5 (dotted line) is characteristic for layers with a higher concentration of unintentional dopants (n $\sim 1.10^{16}$ cm^{-3}), as obtained for growth temperatures T $\geq$ 700°C (sample 2). The exciton line is no longer resolved and the DA and BA lines have merged into a single band of 6.4 meV width. It peaks at 1.492 eV and is due to carbon acceptors. From these measurements, we conclude that in our MO-VPE layers two residual acceptors are detected by P.L. : zinc when grown at low temperatures and carbon when grown at higher temperatures. The non-observation of zinc at high growth temperatures can be explained by the fact that the incorporation of zinc during MO VPE strongly decreases with increasing temperature.

The identification of zinc in MO-VPE GaAs has not been reported previously. Stringfellow (30) only identified carbon, his spectra being similar to the one of sample 2. Seki et al. (2), from a spectrum similar to that of sample 1 concluded that silicon was present as an acceptor in their layers.

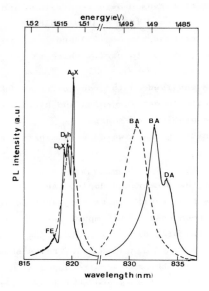

FIGURE 5 : 4 K photoluminescence spectra for MO-VPE GaAs layers
           sample 1 - full line. Growth temperature 680°, n = 7.10^{13} cm^{-3}
           sample 2 - dotted line. "   "   700°, n = 1.10^{16} cm^{-3}

However, according to the detailed investigation of Ashen et al. (31), zinc
and not silicon, is the acceptor involved.

Systematic studies of the background doping as a function of the origin
of the source materials has been performed by Dapkus (32) and Hess (33). They
show very large variations of the 77°K mobility which can vary from 20 000 cm^2/
V.s to almost 90 000 cm^2/V.s depending upon the suppliers. Our own experiments
confirm quite well this observation, as mentioned in table 2. Note also that
TEGa gives rather disappointing results although it has been found purer than
TMGa (2 , 27).

TMGa source	AsH$_3$ source	$N_D - N_A$ cm^{-3}	μ_{77} cm^2/V.s	N_D cm^{-3}	N_A cm^{-3}
S	M 1	3. 10^{14}	68 000	8.10^{14}	5.10^{14}
AV	M 1	2. 10^{14}	88 000	6.10^{14}	4.10^{14}
L	P 1	1. 10^{14}	90 000	5.10^{14}	4.10^{14}
TEGa source					
AV	P 2	1. 10^{15}	49 000	2.10^{15}	1.10^{15}

TABLE 2

Arrangements were made with one supplier to test his most recent production, a three time increase of the 77°K mobility was obtained compared to the previous batch.

Further improvement of u_{77} is obtained by fractional distillation of the TMGa materials with a maximum value of u_{77} = 125 000 cm^2/V.s (33). Tentative assignment suggest that the residual donor could be carbon and that the shallow acceptors are C and Zn in varying relative concentrations.

The growth of high quality $Al_x Ga_{1-x} As$ raises the same purity problems but adds an even more severe risk of contamination by oxygen and water vapour.

Stringfellow (28) pointed out that if 10 % of the oxygen passing through the reactor with an oxygen partial pressure of 10^{-6} were incorporated into the solid $Al_x Ga_{1-x} As$, it would produce a doping level of more than 10^{20} oxygen/cm^3 for normal growth conditions. He has proposed a gettering technique using graphite baffles to catalyze the reaction of TMAl with residual O_2 and H_2O to form $Al_2 O_3$.

This device is shown in figure 2. Similar gettering effect has been shown to occur on the walls of the horizontal growth chamber used in our work on $Al_x Ga_{1-x} As$ (34). The main source of oxygen and water seems to be the arsine cylinder.

The photoluminescence efficiency of $Al_x Ga_{1-x} As$ depends critically upon the oxygen (or water) gettering (28) and for steady state growth conditions it also increases with increasing growth temperature (35). 4 K photoluminescence has been used to identify some shallow acceptors in high quality layers. Spectra for various $Al_x Ga_{1-x} As$ layers are shown in figure 6. These spectra were obtained from samples grown in systems in which care was taken to reduce the residual amount of water vapour and oxygen.(38).

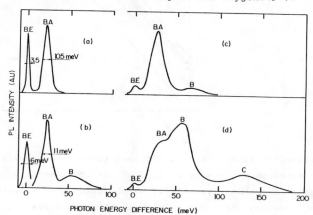

FIGURE 6 : 4 K photoluminescence spectra of four $Al_x Ga_{1-x} As$ MO-VPE layers
a)   x = 0.146, growth temperature T = 800°C, n = $1.7.10^{16}$ undoped
b)   x = 0.195,   "   "   T = 800°C, n = 1.10^{16} undoped
c)   x = 0.308,   "   "   T = 800°C, n = $2.5.10^{17}$ Se doped
d)   x = 0.29,   "   "   T = 700°C, n = 9.10^{16} Se doped

For samples a, b, c, the growth conditions clearly were more fully optimized than for sample d, which was grown at 700°C. Note, however, that the
optimization depends on the growth system involved. Layers with good PL intensities at 300 K and 4 K showing at 4 K only BE and BA luminescence were grown
at 700°C in a system where water of oxygen contamination was extremely reduced.
The best spectra show only two lines. Similar to GaAs, the high energy line
is dominantly due to an exciton bound to a neutral acceptor, the other one to
a BA transition.

Notably in the samples grown at the lower temperatures or with higher
aluminium content, the BE line is rather weak and additional bands are observed. These are labelled B and C in figure 6.

A survey of all peak energies observed as a function of alloy composition
is presented in figure 7. Note that the bound exciton energies have been used
to determine x and thus by definition are on the line representing their
energy dependence on x.

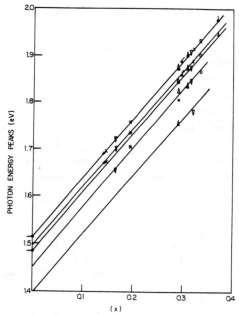

FIGURE 7 :

Photon energy peaks
versus composition
for $Al_xGa_{1-x}As$.

The symbols are
different for each
layer.

First consider the peaks B and C. Extrapolation of the B data to GaAs
leads to an estimated photon energy of 1.45 eV. At this energy a luminescence
band is observed in some MBE layers (36), which could have the same (unknown)
origin. We note, however, that adding Ge during LPE gives rise to a luminescence peak which is similar to our B-line. Peak C can be extrapolated to
1.4 eV in GaAs. This is close to the Mn luminescence peak in GaAs at 1.407 eV
(37). We suggest that Mn is involved in band C in $Al_xGa_{1-x}As$ layers.

The line connecting the shallower BA peaks as a function of x has been calculated using effective mass approximations, similar to ref. 30. This is represented in more detail in figure 8 ; here the theoretically expected difference in photon energy between the BE and BA peaks is plotted as a function of x, together with our experimental data. In addition, the experimental data of ref. 30 are plotted. The acceptor ionization energies of ref.31 were used, that is 26, 28, 30.7 and 34.5 meV for C, Be and Mg, Zn and Si and Cd respectively. From Figure 8, it can be concluded that in most of our epilayers, C is the dominant acceptor. In some layers, grown at lower temperatures, Zn is present. The same conclusions hold for the material grown by Stringfellow (30) ; there is also evidence for Si as an acceptor in one of his layers.

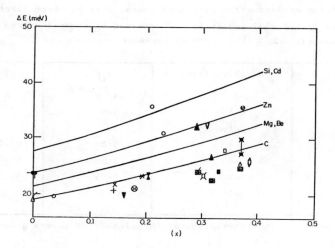

FIGURE 8 : Energy difference E between the BE and BA photon energy versus composition. The full lines are calculated for carbon, magnesium and beryllium, zinc, silicon and cadmium. The symbols are different for each epilayer.

4 DEVICE ORIENTED GROWTH

A large number of devices fabricated with MO-VPE active layers has been reported. The range of applications goes from microwaves to optoelectronics but it is certainly in the field of lasers that most of the interest has been focused. Things are moving very fast and each issue of the technical reviews brings several new results.

GaAs - $Al_x Ga_{1-x}$ As DH injection lasers are essentially identical for all three growth techniques, namely LPE, MBE and MO VPE, with a threshold current density of about 1 000 Å/cm^2 for a GaAs active layer of about 0.20 μm thick (28). In addition, shorter wavelength lasers emitting in the range 725 - 820 nm have also been obtained with $Al_x Ga_{1-x}$As active region. Mori (39) recently reported a 780 nm laser (MO - VPE) superior to the previous LPE ones with threshold current density values as above and life times in excess of 5 000 hours.

For the longer wavelength lasers made of quarternary GaInAsP lattice matched to InP, a lot of work has been performed using the low pressure technology. Table 3 summarizes the state of the art (40).

In view of future devices, the growth of quantum wells is probably the most interesting achievement of the last years. This has been first reported by Dupuis and Dapkus (41) and it is now obtained by several groups using either atmospheric (42, 43) or reduced pressure reactors (44).

Wavelength μm	d (active layer) μm	J_{th} kA/cm^2	T_o K
1.15	0.4	5.9	
	pulsed		
1.22	0.4	1.2	65-70
1.27	0.2	1.5	65-80
1.29	0.2	1.05	65-75
1 5	0.48	2.5	

TABLE 3 : Summary of characteristics of DH GaInAsP/InP lasers grown by LP MO CVD (after 40).

Recently, very low threshold current densities have been reported for multiple quantum well lasers with graded index separate confinement grown by MBE (45). This has motivated two groups involved in MO VPE (46,47) which succeeded both to obtain threshold current densities of about 250 A/cm^2. The structure investigated by Kamenset (46) is given in figure 9, the structure reported by Hersee (47) is similar except that the graded index $Al_x Ga_{1-x}$As layer is constitued of discrete steps.

It is worth discussing what abruptness in change of Al content in quantum wells can be expected in MO VPE with the help of data obtained by Frijlink and Maluenda (43).

They used $Ga(CH_3)_3$, $Al(CH_3)_3$, AsH_3 and H_2 as a carrier gas. The growth temperature was 650°C, the reactor pressure 1 atm., the growth rate 5 Å/sec. The wells were made in continuous growth. The gas transport in the reactor was arranged so as to be able to change the gas composition over the wafer in a controlled way within 0.1 sec.

They have prepared an epitaxial structure with 25 Å wells as shown in figure 10 and measured the photoluminescence. The sample excited at 4 K with a CW argon laser at 5145 Å. Figure 10 shows the obtained spectrum. The width of the peak is 10.5 mV. No luminescence is traced in the region between the peak and the substrate luminescence showing that no alloy clustering is present.

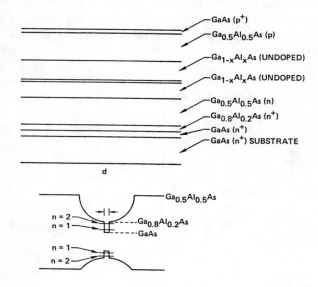

GaAs (p$^+$)
Ga$_{0.5}$Al$_{0.5}$As (p)
Ga$_{1-x}$Al$_x$As (UNDOPED)
Ga$_{1-x}$Al$_x$As (UNDOPED)
Ga$_{0.5}$Al$_{0.5}$As (n)
Ga$_{0.8}$Al$_{0.2}$As (n$^+$)
GaAs (n$^+$)
GaAs (n$^+$) SUBSTRATE

FIGURE 9 :

Schematic diagram of the modified quantum well laser and the associated energy band diagram

d

Ga$_{0.5}$Al$_{0.5}$As
n = 2
n = 1
Ga$_{0.8}$Al$_{0.2}$As
GaAs
n = 1
n = 2

FIGURE 10 :

Photoluminescence spectrum of two 25 Å GaAs quantum wells separated by a 40 Å thick Ga$_{0.46}$Al$_{0.54}$As barrier and with 1 um thick ternary layers on either side. The sample temperature was 4 K.

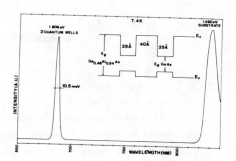

In order to estimate the abruptness in change of Al-content at the interfaces, they also have grown a layer of $Ga_{1-x}Al_xAs$ containing four GaAs layers of 30 Å, 45 Å, 70 Å and 100 Å thickness (figure 11). The Al mole fraction x was found by double X-ray diffraction to be 0.54. The thicknesses of the layers are deduced from steady state growth speed determined by step measurements on thicker layers of GaAs and GaAlAs and from SIMS depth profiling.

Figure 11 shows the measured luminescence spectrum of the four well sample. The arrows indicate the places where the peaks should be according to the calculated relation between wavelength and well width corresponding to the n = 1 electron to heavy hole transition energy for a perfectly rectangular well.

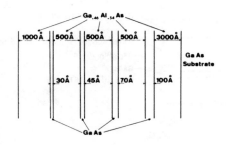

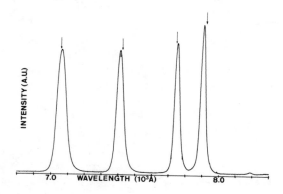

FIGURE 11 : The upper part shows four GaAs wells of different sizes grown in one sample.

The lower part gives the photoluminescence spectrum of the heterostructure measured at 4 K. The arrows indicate the places where the peaks should be according to the calculation.

If the Al-content were to change gradually at the interfaces, the wavelength versus well width relation would be modified. To estimate the size of this modification in the case of the four wells of 30, 45, 70 and 100 Å, a simple model was chosen in which the Al content as a function of its z position is described by :

$$x' = \begin{cases} x & z \leqslant 0 \\ x \, e^{-\frac{z}{L_t}} & 0 < z < d \\ x \, (1 - e^{-\frac{z-d}{L_t}}) & z \geqslant d \end{cases}$$

As L_t increases, the luminescence peak shifts to a shorter wavelength for a given well width d.

Figure 12 shows the result of a calculation which made use of direct numerical integration of the Schrödinger equation to find the real (unperturbed) states). The shift for a narrow well appears to be much larger than for a large well. Comparing figures (11) and (12) one finds that Lt is smaller than 5 Å.

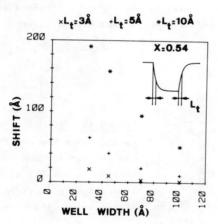

FIGURE 12 :

Luminescence peak shift to shorter wavelength which has to be subtracted from the emission wavelength in the case of a non-abrupt transition in Al content at the interfaces. The drawing shows the assumed exponential composition profile with characteristic transition width L_t.

The other interesting application of the ultra-thin layers is the modulation doped heterostructures.

Considerable mobility enhancement at a GaAs - $Ga_{0.7}Al_{0.3}As$ interface has already been obtained by Coleman (49). Using different optimized growth temperatures for the growth of GaAs and GaAlAs layers, resp. 615°C and 750°C, and selenium as a dopant, they achieved a Hall mobility of 45 000 cm^2/Vs at 77 K.

In the work of Maluenda and Frijlink (48), n-doping of the GaAlAs was realized by introduction of SiH_4 in the gas mixture. Single accumulation layer structures were made, with a 1500 Å thick $Ga_{0.7}Al_{0.3}As$ layer doped at about

$7.5.10^{17}$ at/cm^3 Si except for a 80 Å thick non intentionally doped $Ga_{0.7}$ $Al_{0.3}As$ next to the interface with the 0.5 um thick non-intentionally doped GaAs layer underneath. They used a semi-insulating Cr-doped substrate, oriented (100) tilted 6° off. The reactor and the growth conditions were the same as those used for the quantumwells.

Hall measurements were done by the van der Pauw method, using clover shaped samples in a magnetic field of 1800 Gauss. The 77 K Hall mobilities were found to be in the 70 000 - 80 000 Cm^2/V.s. region, the best wafer exhibiting a Hall mobility of 6 700 cm^2/V.s. at 300 K and 80 000 cm^2/V.s. at 77 K with a mobile carrier concentration per unit area of 8.3 10^{11}cm^2 at 77.K, deduced from the Hall measurement.

Figure 13 shows the influence of the thickness of the undoped $Al_xGa_{1-x}As$ layer, so-called the "spacer", upon the 77°K mobility and the sheet carrier concentration. The mobility peaks for a spacer thickness of about 100 Å, whereas the carrier concentration decreases with increasing spacer thickness because less carriers are injected.

Similar two-dimensional electron gas also occurs at the interface between InP and $Ga_{0.47}In_{0.53}As$. This was first reported by Razeghi (50) who has obtained a 77°K mobility of 30 000 cm^2/V.s. for an equivalent carrier concentration of 2.10^{16} cm^{-3} in a periodic structure of 5 wells. Each InP and GaInAs layer was 250 Å thick.

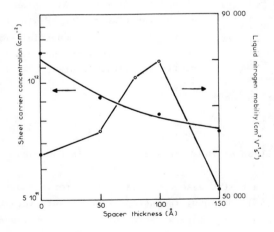

FIGURE 13 :

Variation of the mobility and the carrier concentration in the 2 D electron gas as a function of the spacer thickness.

In the field of microwave devices, MO-VPE has already been used to grow materials for MESFET's and heterostructures FETs. This requires to control successively the low carrier concentration for the buffer layer and the doping uniformity for the active layer. The performances reported by Nakanasi(51, 52) are well comparable with the best chlorine-VPE results. The uniformity is very critical, especially for large devices like power MESFET's fabricated using the recessed gate technology.

Experimental data, obtained by André (53) on such materials are shown in figure 14. A C(V) measurement set-up plots the reverse bias versus carrier concentration or layer thickness. This plot gives also the pinch-off voltage V_P which corresponds to the complete depletion of the active layer. In figure 14, V_P is about 5.2 Volts and the variation is $\Delta V_P = \pm 0.2$ Volts.

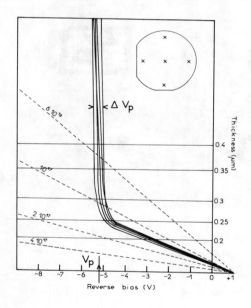

FIGURE 14 :
Variation of the pinch-off voltage of MESFET's layers.

Several heterostructure FET materials have also been obtained. They are

i) the heterobuffer layer FET which uses $Al_x Ga_{1-x} As$ (54),

ii) the heterojonction FET (HJ FET) based on p. $Al_x Ga_{1-x} As$ grown on n GaAs (55)

iii) the MISFET which is obtained by growing insulating $Al_x Ga_{1-x} As$ on n-GaAs (53).

The two first devices had some advantages but were, at the time they have been investigated, rather limited by the quality of the interfaces.

The recent work on MISFET's benefits of the progress of the technology and looks more promising. This work will be published elsewhere.

It is difficult to be exhaustive in this field of device oriented growth where interesting work is also carried on for LED's (56), solar cells and GaAs photocathodes. For this last device, where minority carrier diffusion length and surface specularity are key parameters, MO-VPE seems to do the suitable technique (57). Figure 15 shows a typical photocathode material which consists of a GaAs-GaAlAs double heterostructure, about 15 μm.thick grown on a 2 inch GaAs wafer. The density of macroscopic defects is largely less than 0.5 cm^{-2}.

FIGURE 15 : Photograph of an epitaxial wafer for
GaAs photocathodes.

5 CONCLUSION

This review intended to demonstrate that MO-VPE has reached the level of maturity which allows the fabrication of devices. In fact some weaker aspects of the technique have also been underlined.

The first problem is the availability of pure starting materials. The improvement in quality is in progress and some suppliers offer now a new upgraded quality. Nevertheless sophisticated distillation and/or purer preparation methods would be necessary to provide pure and abundant metalorganic sources. Progress in the hydride AsH_3, PH_3 would also be interesting in view of reducing the oxygen and water pollution.

The second problem deals with toxicity and fire risks. This may become serious if the MO-VPE equipments were run by unexperienced people. On the other hand the VPE production of $GaAs_{1-x}P_x$ LED's already uses arsine and phosphine and the safety rules are fixed. One definite advantage of the use of adducts is their stability at room temperature, which makes them easy to transport. The actual disavantage is the novelty which brings us back to the first point.

Whereas the growth of GaAs - $Al_xGa_{1-x}As$ looks well established and successfully performed in many laboratories, only a few groups are able to grow good quality InP and related quaternaries. Furthermore, technical routes are under discussion with the choice between on one side adduct precursors and low pressure + phosphine pyrolysis on the other side. All the other compounds which were proved feasible are at a much less advanced stage. It is nevertheless worth noticing that no other epitaxy technique but LPE has been able to produce the complete range of compounds.

From figure 1, it is obvious that MBE and MO-VPE are competing and this has been illustrated in the case of quantum well lasers. Both techniques are still behind LPE for producing high quality $Al_xGa_{1-x}As$ with x $\leqslant$ 0.25 and in both cases, the electronic properties improvement involves an increase of the growth temperature. Tentatively, this could be related to carbon contamination with Al-C bands which are much more stable than the Ga-C ones. Another common point could be the growth mechanism if the hypothesis given in section 2 is correct. The competition seems then more related to people passion in laboratories which can afford both techniques. At the industrial level, it will only be the investments and the producing capacity which will be taken in account.

Among the many devices reported here, two important markets can be forseen for the next future, they are the DH lasers and the FET's. State of the art results in research laboratories are sufficiently convincing for reasonable start of development and medium scale production.

Acknowledgements

I gratefully acknowledge R. BHAT from Bell Telephone Laboratories, P.D. DAPKUS from Rockwell International, J.P. DUCHEMIN from LCR Thomson CSF, R. MOON from Hewlett-Packard, R.H. MOSS from British Telecom Research Laboratories, for the supply of their recent preprints. Thands are due to J.P. ANDRE, P. FRIJLINK and J. MALUENDA from LEP and to A. BRIERE and P. FAUVEL for technical assistance. Finally I would like to thank A. MIRCEA-ROUSSEL for critical discussion of the manuscript.

REFERENCES

(1) H.M. MANASEVIT, J. Cryst. Growth, 13/14, (1972), 306

(2) Y. SEKI, K. TANNO, K. IIDA and E. ICHIKI, J. Electrochem. Soc., 122, (1975) 1108

(3) S.J. BASS, J. Cryst. Growth, 31, (1975), 172

(4) J.P. ANDRE, A. GALLAIS and J. HALLAIS, Inst. Phys. Conf. Ser. n° 33a, (Institute of Physics, London, 1977), 1

(5) R.D. DUPUIS and P.D. DAPKUS, Appl. Phys. Lett., 31, (1977), 466

(6) J.P. DUCHEMIN, M. BONNET and . HUYGHE, Revue Technique Thomson-CSF, 9, (1977), 685

(7) N. HOLONYAK Jr., R.M. KOLBAS, W. LAIDIG, B.A. VOJAK, R.D. DUPUIS and P.D. DAPKUS, Inst. Of Phys. Conf. Ser. n° 45, (Institute of Physics, London 1979), 387

(8) J.P. DUCHEMIN, M. BONNET, C. BENCHET and F. KOELSCH, Inst. Of Phys. Conf. Ser. N° 45, Institute of Physics, London, 1979, 10

(9) J.P. HIRTZ, J.P. DUCHEMIN, P. HIRTZ, B. DE CREMOUS, Electron. Lett., 16 (1980), 275

(10) MR. LEYS AND H. VEENVLIET, J. Cryst. Growth, 55, (1981), 145,

(11) L. HOLLAN, J. HALLAIS and J.C. BRICE, Current Topics in Material Science vol. 5, North Holland 1980, p. 1

(12) D.J. SCHLYER AND M.A. RING, J. Electrochem. Soc. 124, (1977), 569

(13) C.T. FOXON, Acta Electron., 21, (1978), 139

(14) H.M. MANASEVIT, Journal of Cryst. Growth 55, (1981), 1-9

(15) A.K. CHATTERJEE, M.M. FAKTOR, R.H. MOSS and E.A.D. WHITE to be published in the Proceedings of the "Colloque sur l'Epitaxie des Semiconducteurs, Perpignan 1982"

(16) J.P. ANDRE, unpublished results

(17) C.B. III COOPER, M.J. LUDOWISE, V. AEBI and R.L. MOON, Electron. Lett., 16, 1, (1980), 20-21

(18) K.W. BENZ, H. RENZ, J. WIEDLEIN and M.H. PILKUHN, J. Electron. Mater. 10, 1981, 185-192

(19) M. RAZEGHI, M.A. POISSON, J.P. LARIVAIN, J.P. DUCHEMIN, to be published in Journal of Electronic Materials.

(20) A. ZAOUK, E. SALVETAT, J. SAKAYA, F. MAURY and G. CONSTANT Journal of Cryst. Growth, 55, (1981), 135

(21) W.T. DIETZE, M.J. LUDOWISE, C.B. COOPER, Electron. Let. 17, (1981), n° 19, 698

(22) S.J.C. IRVINE and J.B. MULLIN, Journal of Cryst. Growth, 55, (1981), 107-115

(23) METALORGANIC VAPOR PHASE EPITAXY, Journal of Cryst. Growth, 55, (1981)

(24) C.R. LEWIS, W.T. DIETZE, M.J. LUDOWISE, Electron. Lett., 18, n° 13 (1982), 569

(25) R. MELLET, R. AZOULAY, L. DUGRAND, E.V.K. RAO, A. MIRCEA, Inst. Phys. Conf.
Ser. n° 63, (Institute of Physics London, 1982), 583

(26) J. HALLAIS, J.P. ANDRE, P. BAUDET and D. BOCCON-GIBOD, Inst. Phys. Conf.
Ser. n° 45, (Institute of Physics London (1979), 361

(27) R. BHAT and V.G.KERAMIDAS, SPIE Conference in Los Angeles, C.A. 1982

(28) G.B. STRINGFELLOW, J. Cryst. Growth, 53, (1981), 42

(29) B.W. WILLIAMS and H.B. BEBB Jr., Semiconductors and Semimetals, 8,
(Academic Press, New-York, 1972), 371.

(30) C.B. STRINGFELLOW and R. LINNEBACH, J. Appl. Phys. 51, (1980), 2212

(31) D.J. ASHEN, P.J. DEAN, D.T.J. HURLE, J.B. MULLIN and A.M. WHITE,
J. Phys. Chem. Solids, 36, (1975), 1041

(32) P.D. DAPKUS, H.M. MANASEVIT, K.L. HESS, T.S. LOW and G.E. STILLMAN
J. Cryst. Growth, 55, (1981), 10

(33) K.L. HESS, P.D. DAPKUS, H.M. MANASEVIT, T.S. LOW, B.J. SKROMME and
G.E. STILLMAN, to be published in J. of Electrochem. Materizls

(34) G. LAURENCE, F. HOTTIER and J. HALLAIS, J. Crystal growth, 55, (1981)
198

(35) J. HALLAIS, J.P. ANDRE, A. MIRCEA-ROUSSEL, Journal of Electronics Mat.
10, (1981), 665

(36) A. MIRCEA-ROUSSEL, unpublished results

(37) T.C. LEE and W.W. ANDERSON, Solid State Comm., 2, (1964), 265

(38) A. MIRCEA-ROUSSEL, A. BRIERE, J. HALLAIS, A.T. VINK, and H. VEENVLIET,
J. Appl. Phys. (1982)

(39) Y. MORI, to be published in the Proceedings of the "Colloque sur
l'épitaxie des semiconducteurs", Perpignan, 1982.

(40) M. RAZEGHI,J.P. HIRTZ, P. HIRTZ, J.P. LARIVAIN, R. BONDEAU, B. DE CREMOUX,
J.P.DUCHEMIN, Electron. Lett., 17, (1981), 597

(41) R.D. DUPUIS, P.D. DAPKUS, H. HOLONYAK, E.A. REZEK and R. CHIN, Appl.
Phys. Lett. 32, (1978), 295

(42) D.R. SCIFRES, R.D. BURNHAM and W. STREIFER, Appl. Phys. Lett. 41, (1982),
118

(43) P. FRIJLINK and J. MALUENDA, Submitted to Jap. J. Appl. Phys. Let.

(44) M. RAZEGHI, N.A. POISSON, J.P. LARIVAIN, B. DE CREMOUS, J.P. DUCHEMIN,
Electron. Lett., 18, (1982), 339

(45) W. I. TSANG, Appl. Phys. Lett. 40, (1982), 217

(46) D. KASEMSET, C.S. HONG, N.B. PATEL and P.D. DAPKUS, to be published in
Appl. Phys. Lett.

(47) S. HERSEE, M. BALDY, P. ASSENAT, B. DE CREMOUX, J.P. DUCHEMIN, Electron.
Lett., 18, (1982), 618.

(48) J. MALUENDA and P. FRIJLINK, to be published.

(49) J.J. COLEMAN, P.D. DAPKUS and J.J.J. YANG, Electron. Lett., 7, (1981), 66

(50) M. RAZEGHI, M.A. POISSON, J.P. LARIVAIN, B. DE CREMOUX and J.P. DUCHEMIN
Electron. Lett., 18, (1982), 339

(51) T. NAKANISI, T. UDAGAWA, A. TANAKA and K. KAWAI, J. Cryst. Growth, 55, (1981), 255

(52) T. SHINO, S. YANAGAWA, Y. YAMADA, K. ARAI, K. KAMEI, T. CHIGIRA and T. NAKANISI, Electron. Lett., 17, (1981), 738

(53) J.P. ANDRE, C. SCHILLER, AMITONNEAU. A. BRIERE and J.Y. AUPIED, to be published in the Proceedings of GaAs and Related Compounds, Albuquerque (1982)

(54) J. HALLAIS. J.P. ANDRE P..BAUDET and D. BOCCON-GIBOD, Inst. Phys. Conf. Ser. n° 45, (1979), 361-370

(55) J. HALLAIS and D. BOCCON-GIBOD, Acta Electronica, 23, (1980), 339-345

(56) M.J. TSAI, M. TASHIMA, B. TWU and R.L. MOON, to be published in the proceedings of GaAs and Related Compounds, Albuquerque, 1982

(57) J.P. ANDRE, P. GUITTARD, J. HALLAIS, and C. PIAGET, J. Cryst. Growth 55, (1981), 235.

The Sensor Lag: A Threat to the Electronics Industry?

S. Middelhoek

Department of Electrical Engineering, Delft University of Technology
P.O. Box 5031, 2600 GA DELFT, The Netherlands

SUMMARY

The penetration of sophisticated micro-electronic circuits into
traditionally non-electronic products has been seriously hampered by
the lack of low-cost, reliable transducers for input and output
functions.   Because the traditional electronics markets show serious
market saturation, only the market for traditionally non-electronic
products constitutes a major growth possibility for the electronics
industry. In the paper the different generic technologies for making
sensors are briefly discussed. Silicon planar technoloy appears to be
very suitable for the construction of sensors that are compatible with
micro-electronic components.   For each of the five signal domains
preliminary results on a new silicon sensor are presented.

1 INTRODUCTION

The revolutionary development of micro-electronics over the last
two decades has been celebrated in many places and on many occasions.
There is general agreement that the astronomical increase in the
performance/price ratio of micro-electronic components has given an
important impetus to what is commonly called the "information

society", one in which information and information-processing systems
play a major role.

Information-processing systems come in numerous shapes and comprise
not only equipment such as computers, calculators, automatic landing
systems, process controllers , etc. but also products such as scales,
word processors, sewing machines, thermometers, compasses, burglar
alarms, hearing aids, talking dolls and, in fact, all equipment with
which, in order to obtain a result, some information is being
processed.

Fig.1   Functional block diagram of an
information processing system.

Without exception an information-processing system consists of
three units, as shown in Fig.1 (1). In the input transducer, often
called sensor, a measurand such as temperature, pressure,
displacement, magnetic field direction or humidity is converted into
an electrical signal. In the signal processor in the center of the
system the electrical signal is in some way modified, that is,
amplified, filtered or, say, converted from an analogue into a digital
signal. In the output transducer the electrical signal is converted
into a signal which:

1. can be perceived by one of our senses (display),
2. can cause some action (actuator),
3. can be stored (memory device) or
4. can be transmitted to another location (transmitter).

The fact that signal conversion in a transducer is always based on
energy conversion makes it convenient to distinguish among six signal
domains:

1. radiant (light, X-rays, γ-rays),
2. mechanical (pressure, flow, level),
3. thermal (temperature, heat),
4. electrical (current, dielectric constant),

5. magnetic (field strength, flux density) and
6. chemical (pH, humidity).

Conversion from one of the non-electrical signal domains to the electrical signal domain takes place in the input transducer or sensor. Conversion from the electrical to one of the five non-electrical signal domains occurs in the output transducer, whereas in the center at the in- and output of the signal processor the same electrical signal domain is observed (Fig.2), so that no signal conversion takes place in the signal processor (2).

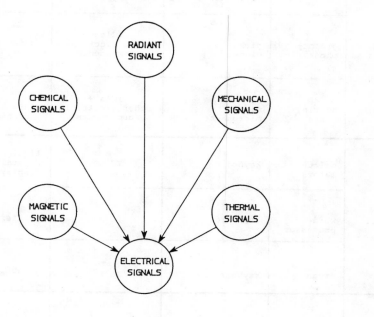

Fig.2   Diagram representing the five
possible signal conversions in transducers.

Based on this analysis of information-processing systems, three groups of industrial products in which micro-electronic components are or can be used can be distinguished (Table 1):

I.   The first group encompasses the traditional electronic products, in which electronic components have already been incorporated for many years. This group mainly consists of television sets, radio receivers, audio-equipment, instruments and computers. The development of these products occurred rather gradually. In the

TABLE 1		Product groups and transducers		
group	product	input transducer	signal processor	output transducer
I	TV set	dipole antenna	IC's + other electronic components	cathode ray tube
I	grammo-phone	pickup-element	IC's + other electronic components	loudspeaker
I	hearing-aid	mini-microphone	IC's + other electronic components	mini-loudspeaker
II	calcu-lators	keyboard	IC	liquid - crystal - display
II	word-processor	keyboard	IC's	printer
II	organ	keyboard	IC's	loudspeaker
III	letter scale	weight sensor	IC	liquid - crystal - display
III	CO-guarded bath heater	CO-detector	IC	pump switch
III	food poisoning indicator	bacteria sensor	IC	red and green LED

past five decades vacuum tubes, relays and passive components have successively been replaced in these products by transistors and by respectively small-scale, medium-scale, large-scale and very large-scale integrated circuits. The replacement of vacuum tubes by micro-electronic components had a very positive influence on the performance/price ratio of these products. The in- and output transducers required for the development of the above products, such as antennas, phonograph pickup elements, loudspeakers, push buttons, picture tubes, etc. were in time generally made available and showed acceptable performance/price ratios.

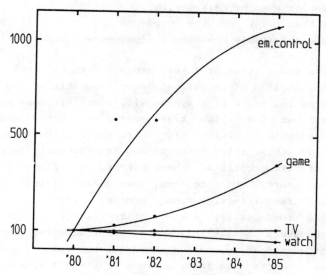

Fig.3   Market development for several
electronic products (From (3)).

Today the demand for these traditional electronic products is not growing anymore and in some sectors (Fig.3) even serious market saturation can already be observed (3). Because many companies and nations are heavily competing for this market, this traditional product market in any case will not allow most companies producing micro-electronic components to survive the coming years much less to increase their production.

II.   The second group consists of products which use more or less the same electronic concepts and components developed for the

products in the first group but in an innovative way. For these
products the allocation of suitable in- and output transducers
did not pose too serious a problem. Such products are electronic
watches, alarms, calculators, toys, organs, games, home
computers, word processors, cash registers, cameras, etc. For
the development of these products well-known and well-developed
transducers such as push buttons, paddles, quartz resonators,
photodiodes, cathode-ray tubes, light-emitting-diode and
liquid-crystal displays, stepping motors, printing heads, etc.
were available. Though many of these product markets still have
future growth potential (Fig.3), some sectors such as chips for
electronic watches, calculators and organs already show signs of
saturation or even of market slump.
It seems that many of the products of group II no longer
constitute a reliable base for a sound growth industry.

III. Finally, a third group of future products can be envisioned for
     which the design of the signal-processing unit in the center of
     the information processing system will certainly pose no serious
     problems, but for which transducers with a performance/price
     ratio comparable to that of micro-electronic circuits do not yet
     exist. To this third group belong such products as electronic
     scales, traffic controls, thermometers, gas-flame regulators,
     smoke detectors, CO detectors, person-identification systems,
     pressure transmitters, flow control systems, electronic
     compasses, food-quality testers, tire-pressure warning systems,
     bike tachometers, health indicators, plant watering systems, etc.
     With not much fantasy one can enumerate hundreds of such not yet
     available products. In fact, a group in the USA apparently came
     up with well over 10,000 products.
     In view of the present and anticipated saturation of the first
     and second product markets, this third group of products can
     constitute, in fact, the major future growth market for the
     electronics components and systems industry and can therefore be
     decisive for the future health of this industry.

For most of the products of group III it is in fact not difficult
to design a signal processor based on existing integrated circuits.
It is also not usually difficult where necessary to design a special
custom integrated circuit. In most cases the display of the
information can be done reasonably well with the help of
liquid-crystal displays, light-emitting diodes or cathode ray tubes.

The performance/price ratio of these displays is often acceptable, though the search for a good flat screen display is still going on. The real problem starts when one tries to allocate proper input transducers or sensors and actuators. No sensors with the desired performance/price ratio are yet available for converting measurands such as nuclear radiation, X-radiation, light-beam direction, weight, level, flow, heat flow, magnetic field, pH, CO concentration or moisture, to name a few.

However, where these and many other sensors become available a large number of new, useful and of course also non-useful products could be realized. The large number of micro-electronic components they employ could lead to an increased turnover in this market. From the above arguments one may postulate, that it is of vital interest to the electronics industry either to develop themselves or to support the development of new transducers, that is to say new sensors, displays and actuators. Or stated more strongly it appears that the present sensor lag might pose or even is a serious threat to the future of many electronics industries.

In view of the importance of the sensor field this paper deals with several topics pertinent to the development of new sensors. In section 2 new technologies suitable for the production of novel sensors will be briefly reviewed. Silicon planar technology is an important contender, as will be shown.

In section 3 a number of new sensor examples which are current subjects of research in the author's laboratory come up for review. In the last section conclusions and a guarded future outlook on the field of sensors will be presented.

2 SENSOR TECHNOLOGIES

2.1 Introduction

For many decades work on transducers has been performed in a large number of small, specialized industries and in the R&D laboratories of large multinational companies. This has led to an immense number of measuring principles and devices. Books describing this field often have encyclopedic dimensions.

With respect to technology the whole transducer field can be represented as is shown in Fig.4 (1). The field contains, besides all kinds of macroscopic principles such as conventional mercury thermometers, bourdon pressure gauges, pneumatic controllers, linear variable differential transformers for displacement measurement, etc., also the important group of solid-state transducers.

TRANSDUCERS

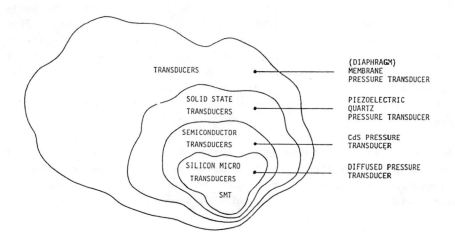

Fig.4   Survey of the transducer technology field.

The operation of these transducers is based upon phenomena occurring in the solid state. This group of solid-state transducers contains not only devices such as piezoelectric quartz pressure transducers, platinum resistance thermometers, LiCl moisture sensors, etc. but also the interesting sub-group of semiconductor transducers. The functioning of these transducers is related to the occurrence of a forbidden energy gap for charge carriers and a filled valence band in the semiconducting materials.

Furthermore, this group of semiconductor transducers contains in addition to InSb Hall plates, CdS photodetectors, GaAs pressure sensors, etc., also the very important group of silicon micro-transducers (4). The operation of these devices is mostly based on the semiconducting properties of silicon. However it appears that silicon can also be used as a mechanical construction material for transducers by exploiting the silicon planar processing technoloy and not the semiconducting properties of silicon per se (5).

In order to combat the sensor lag different approaches are being taken. In one approach the characteristics of already existing and reliable transducers are improved by adding signal-processing circuits (6). Today, for instance, it is not difficult to correct a nonlinear dependence or an offset of a measuring element by suitably programming a microprocessor as long as the nonideal characteristics are reproducible. An unwanted temperature sensitivity can also be easily corrected. The transducer and the integrated circuit are often built together in a hybrid package. When both the transducer and the integrated circuit are mass produced and therefore inexpensive and the combination of both components yields a high-quality transducer, this approach might be highly recommended.

However, very often this solution is not possible, so that another approach to the problem is in order. Often solutions result from the search for new materials or from the innovative use of well-known principles and the transducer field is no exception.

In the following the features of non-silicon generic and silicon technologies for transducers are briefly discussed.

2.2 Non-silicon generic technologies.

In the course of the years a number of technologies for making solid-state transducers has been introduced. Some technologies are rather new, because only now has the processing of certain materials been sufficiently mastered, whereas other materials have already been used for many decades and have been developed to their present state.

Piezoelectric materials

Certain classes of materials show piezoelectric effects. A mechanical strain produces an electric polarization. The inverse effect, i.e. an electrical polarization producing a strain or a dimension variation, will also occur in these materials. Such materials are mainly suited to the construction of transducers in which conversion from the mechanical to the electrical domain occurs, but some other signals also can be detected. Quartz is the most used piezoelectric material, but piezoelectric ceramics or polymers are often used as well. $LiNbO_3$ and PVF_2 are materials that in the last decade have become rather popular for different applications.

Piezoelectric transducers often have the advantage that the measurand is converted into a frequency, which can be measured with great accuracy (7). Because for certain quartz platelets the resonance frequency is a function of the temperature, quartz thermometers can be constructed.

Another device in which the piezoelectric effect can be used is based on surface acoustic wave propagation (8). Because of strain or temperature increases, the path length changes, causing phase changes which can be detected with great accuracy. Some allied crystals show pyroelectricity, which stands for the change of polarization caused by temperature changes. Very sensitive transducers for the measurement of temperature and flow can be made (9).

Polymers

The conductivity, permittivity or mass of certain polymers change slightly when they are inserted in certain gas atmospheres (10). The electrical properties can be measured by constructing sandwich capacitors or by covering interdigital electrode patterns with the polymer. Sensitivity to CO, CO_2 , CH_4 and moisture has been shown succesfully, and it is expected that by carefully studying the electrical proporties of polymers materials will be found in the near future that can be used for atmospheric pollutant detection and humidity measurement.

Metal oxides

Much work is going on in the field of gas-sensitive metal oxides (11). At present the mechanisms are not very well understood, but nonetheless a few materials are on the market with which, under certain circumstances, reasonable sensors can be built. Materials such as SnO_2, ZrO_2, WO_3 and ZnO, with or without catalysts, show sensitivities to H_2, H_2O, O_2, CO, CH_4, etc.

When the understanding and the technology of metal oxides are improved they might provide us with a new class of useful, lowcost gas sensors.

III-V and II-VI semiconductors

Materials such as GaAs, GaP, AlSb, InSb, InAs, CdS, CdSe, ZnO, ZnS, etc. are semiconductors. In most cases, when a semiconductor is needed for the fabrication of a transducer, silicon is used because the technology is very well-known and silicon allows batch fabrication.

However, sometimes silicon does not show the proper physical effects. Therefore, when for instance a direct gap or a piezoelectric semiconductor or a larger temperature range is needed, Si will be replaced by GaAs (12). When a high-mobility material is needed InSb is much better than Si. When a larger bandgap is required CdS may be preferable to Si.

It is also possible to deposit these III-V and II-VI semiconductors on Si substrates to get very interesting devices.

Thick- and thin-film materials

By several techniques thin layers of resistive, dielectric, piezoelectric, semiconducting or magnetic materials can be deposited on suitable substrates. These layers often show the same effect as the bulk material and have the advantage that they can be easily combined with electronic circuits in hybrid packages.

For almost anyone of the signal domains well functioning transducers have been made (13). Thin-film NiCr strain gauges, platinum temperature sensors, capacitive aluminum displacement sensors, thick-film thermistors, thin Ni-Fe film magnetic recording heads., thick-film pH sensors, thin-film thermocouples, etc. are reported in the literature.

Because thin- and thick-films often do not require large processing-equipment investments, these techniques are often suitable for the fabrication of transducers requiring only a small series.

Optical glass fibers

Optical glass fibers have been investigated for use in optical communication systems. It has been proved that transmission along fibers is influenced by a number of perturbations and consequently research was focused on materials and structures that do not show these effects.

One spin-off of this research was that scientists working in the transducer field discovered that glass fibers can also be used to construct very sensitive and convenient transducers. For instance, temperature changes and mechanical perturbations cause polarization and phase changes which can be easily detected. Chemical sensors can also be made, because the optical properties of the cladding influence the transmission properties of the fiber. Many new sensors based on glass fibers can be expected in the future (14).

2.3 Silicon technology

Silicon can be used as amorphous, polycrystalline and monocrystalline material. As the whole integrated-circuit industry is based on the use of monocrystalline silicon, it therefore seems sensible to reap the benefits of previous experience and to also use monocrystalline silicon for the fabrication of transducers (15). Nonetheless it appears that amorphous and polycrystalline silicon are advantageous for certain applications. Their light sensitivity and temperature dependence differ enough from that of monocrystalline silicon to be of use in applications where monocrystalline silicon is not suitable.

However, the majority of known silicon transducers is based on the use of the planar technology and monocrystalline silicon. The use of silicon allows the introduction of well-developed and sophisticated batch-production oriented processing steps and also permits sensors and integrated circuits to be combined on one chip. Such sensors are sometimes called "smart sensors" or "intelligent transducers".

Work on silicon transducers was started many years ago. Silicon has been used for the detection of light since 1960 (16) and an extensive literature on these devices exists. Later on silicon was

used for the measurement of pressure, temperature and magnetic field.
Today a large amount of silicon sensing devices is reported in the
literature.

For nearly all important measurands, sensing devices have been
made.   The most important elements for measuring light intensity,
light images, pressure, acceleration, position, flow, temperature,
temperature difference, magnetic field, pH concentration, gases and
humidity have been recently reviewed by the author (4). In spite of
the large number of silicon sensors investigated the reluctance of the
semiconductor industry to invest in sensor development has led to
a situation where the number of silicon sensors commercially available
at present is much lower than the number of scientifically reported
ones.

3 EXAMPLES OF SOME NEW SILICON MICRO-TRANSDUCERS

In order to merely illustrate the different possibilities of
silicon as a sensing material, some new sensors for which some results
have already been obtianed and which are being studied in the author's
laboratory will be described here. The measurands, as has already
been shown, can be divided into five signal domains. An example of
a sensor will be given for each signal domain.

3.1 Radiant signal domain: strip detector for nuclear particles.

For the performance of experiments with high energy colliding beam
facilities better particle detectors are clearly needed. Two features
are particularly important: 1) improved space resolution and 2) this
on a relatively large area.   Only rather recently did one start to
realize that silicon planar technology might be suitable for realizing
these goals.

An initial interesting approach was reported by Heyne et al. (17),
who fabricated a strip detector consisting of a 8000 Ωcm n-type
silicon substrate covered with 140 µm wide wide strip contacts. These
contacts are obtained by evaporating gold through a foil with
strip-like openings. The gold forms Schottky contacts with the

underlying silicon substrate. In another attempt Kemmer (18) used an implantation and other standard silicon processing steps to obtain p-type strips on 500 Ωcm to 5 kΩcm n-type substrates. Detectors with a microstrip pitch of 200, 50 and 20 μm were fabricated and showed encouraging results.

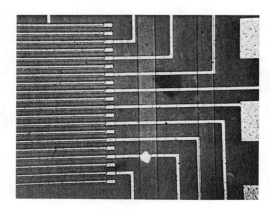

Fig.5   Part of silicon microstrip detector, 10 μm wide p-type strip-electrodes are diffused in a 2 KΩcm n-type substrate.

In our laboratory test structures have been fabricated with a pitch of 20 μm as is shown in Fig.5. The strip pn diodes were obtained by standard p-type diffusion in 2 kΩcm n-type material. The quality of this high-resistivity material was such that after the complete processing sequence the high resistivity of the material was preserved. In preliminary experiments the detectors showed high energy and space resolution as expected.

Silicon planar technology, as is well known, allows the fabrication of strip widths as small as 1 μm, so that in theory the space resolution of the detector can be improved much more. However, because each strip has to be connected to an amplifier, the fan-out problem becomes rather unsolvable. Therefore, innovative multiplexing readout systems have to be conceived in order to fully exploit the advantage of the silicon planar technology for the fabrication of silicon sensors which can detect nuclear particles.

3.2 <u>Mechanical</u> <u>signal</u> <u>domain</u>: <u>position-sensitive</u> <u>detector</u> <u>for</u> <u>the</u> <u>measurement</u> <u>of</u> <u>X-ray</u> <u>and</u> <u>electron-beam</u> <u>positions</u>.

For application fields such as in X-ray diffraction, medical diagnostics and therapy, astromonomy and lithography, a detector for determining the exact position of an X-ray beam is desirable. Such a device could be constructed, by applying on top of an already-developed light-spot position-sensitive detector or PSD (19) a phosphor in which X-rays are converted into photons in the visible spectrum.

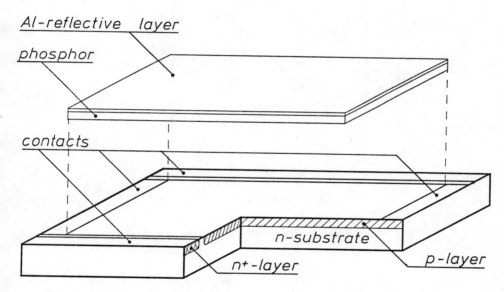

Fig.6   Schematic diagram of the X-ray beam position sensor.

As is shown in Fig.6, the detector consists of a PSD which is obtained by implanting a p-type layer in an n-type epitaxial layer. Both layers have two elongated contacts. When a reverse bias voltage is applied across the pn-junction a light spot creates charge carriers that flow to the nearest contacts. The current distribution appears to give an exact linear relationship with respect to the x- and y-coordinates.

Such a device is very suitable for light beams in the visible spectrum, but because silicon has a very small mass, the device is

practically transparent for X-radiation. In order to make the device
also sensitive to X-ray and electron beams, the PSD is, as shown in
Fig.6 covered by a suitable phosphor.

Several phosphors have been considered. Good results were obtained
by using an aluminium covered Zn (0.2) Cd (0.8) S:AG layer, 100 µm
thick and deposited by thick-film technology. Hardening temperatures
were chosen to be below 180° C.

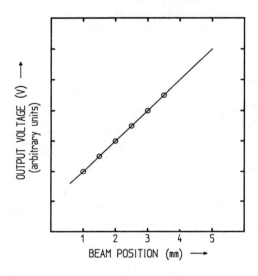

Fig.7   Output voltage in arbitrary units as a
function of the X-ray beam position.

X-rays with energies of 29, 48, 66 and 119 keV were focused on the
detector. The X-rays were converted in the phosphor into a light
spot. The underlying PSD subsequently produced electronic signals
proportional to the coordinates of the light spot. In Fig.7
preliminary results show that the concept is feasible (20). The
resolution for light of the PSD is a few µm. Because the presently
used phosphor is rather grainy (5-10 µm) this resolution is not
obtained for the position of the X-ray beams.

The same device was also used to detect the position of an electron
beam in an electron microscope. Here as well a straight-line
relationship was obtained.

3.3 Thermal signal domain: silicon thermopile detector.

Thermocouples based on the Seebeck effect are used to measure temperature differences or to convert thermal energy into electrical energy. In order to obtain a satisfactory sensitivity, the use of a pile of thermocouples connected in series is often recommended. However, the construction of such a thermopile is rather labor-intensive. Therefore, it makes sense to apply silicon planar technology, with its batch-fabrication features, to the fabrication of thermopiles.

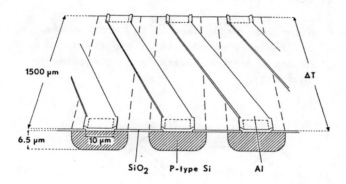

Fig.8   Schematic representation of a part of
        the aluminium p-type silicon thermopile.

In a feasibility study a thermopile was constructed (Fig.8) consisting of 152 series-connected aluminium p-type Si thermocouples filling an area of 3.8 x 1.5 mm^2 (21). The strips were 1.5 mm long. The diffusion depth was chosen to be 6.5 µm and the diffusion-mask-opening width was 10 µm with a spacing of 15 µm.

A representative thermopile gave a 76 mV per degree temperature difference between the opposite edges of the thermopile chip. The thermopile had a total internal resistance of 250 kΩ and a maximum current of 0.3 µA per degree temperature difference. Devices with a larger number of thermocouples, using both p-type and n-type diffusions, might result in even more sensitive devices.

3.4 <u>Magnetic</u> <u>signal</u> <u>domain</u>:   <u>magnetic-field</u> <u>vector</u> <u>sensor</u>.

The best-known silicon magnetic-field sensor is based on the Hall effect.   Another device, called the "magnistor", employs the Lorentz force to deflect the majority charge carriers.   The device consists of one elongated emitter and base structure and two collectors at both sides.   The magnetic field is applied in the   plane of the device. Because of the Lorentz force the distribution of the currents between the collectors is affected.   The difference current appears to be exactly linear to the magnetic-field component parallel to the emitter for a very wide field range.

Using two such devices, perpendicular to each other, allows the measurement of the x- and y-components of the field. However, a more elegant device, occurs when both devices are integrated in one device as is shown in Fig.9 (22).

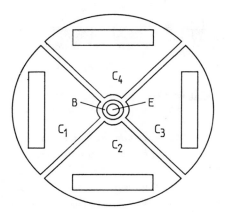

Fig.9   Schematic diagram of magnetic field vector sensor.

The four-collector device consists of four buried n^+sectors with four collector contacts and only one circular emitter and base structure in the center. With this structure the x- and y-components of the magnetic field can be simultaneously detected. Combining a permanent magnet with the sensor makes it possible to construct an angle or rotation detector. In principle, the device is also suitable for the construction of an electronic compass.   However, for this

purpose the sensitivity of the device has to be improved by an order
of magnitude.

3.5 Chemical signal domain: gas sensor.

It is known that the permittivity and conductance of a polymer such
as polyphenylacetylene (PPA) slightly changes upon the absorption of
gases or moisture. This property can be used in a gas sensor. Such
a device consists of a substrate on top of which is situated
a structure made up of 40 pairs of 3500 Å thick aluminium electrodes
filling an area of 1600 µm x 1200 µm (23). Each electrode is 10 µm
wide and 1200 µm long, and the gap between the electrodes is also
10 µm. A 0.77 µm thick polymer (PPA) is spin-coated on top of this
structure. The resulting capacitance has a value of around 3pF.

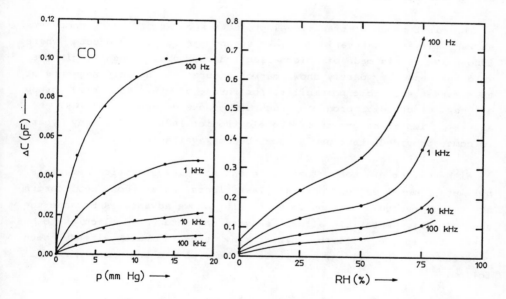

Fig.10   Capacitance change as a function of
CO-pressure and relative humiditiy.

In a measurement sequence a vacuum vessel, containing the sensor,
is evacuated with a rotary pump. In a second step the gas is
admitted. The capacitance of the cell is measured before and after
the admission of the gas. The pressure of the gas is increased in
steps and the capacitance change with respect to the vacuum state is

plotted.   The capacitances are measured in a frequency range between 100 Hz and 100 kHz. Measurements were performed for CO, CO_2, CH_4 and humidity.

Fig.10 shows the results for CO and those for relative humidity. These results indicate that the monitoring of the capacitance of a polymer-covered interdigital capacitor might be useful for the detection of gases and moisture.

4 CONCLUSIONS AND FUTURE OUTLOOK

Three groups of products exist in which micro-electronic components are or can be used.

The first group of traditional products such as TV and computers shows   market   saturation.   The second group of products where innovative use is made of micro-electronics, e.g.   cash registers, word processors, partly shows market saturation and only segments of this market show some possibility for future growth. The third group of new electronic products such as smoke detectors, scales, etc. allows an important growth of the electronics industry for many years to come, provided, that new sensors become available.

With different technologies, such as piezoelectric materials, polymers, metal oxides, optical glass fibers and silicon, new sensors are being deviced. Silicon in particular shows advantages because the use of this material allows sensor and integrated circuits to be integrated in one silicon chip. Many sensors have already been constructed with silicon, and it is to be expected that many more will follow.

Because the availability of sensors is a decisive factor for the future growth of the electronics industry, increased research effort with regard to sensors is being contemplated by many industries. Three approaches can be distinguished.

Large semiconductor industries will concentrate on transducers which will guarantee a low-cost mass market, where the potential of batch fabrication can be fully exploited.

The second group of industries to develop and manufacture transducers are industries of the instrument sector. In order to improve their measurement and control equipment new sensors are required. Such transducers will inevitably be made in small series, will have special features and be of high quality and consequently will show high cost.

The third group of industries that might evolve will, just as in the integrated-circuits field, concentrate on customdesign transducers. Because the production of transducers often requires elaborate technological facilities, this solution might be an efficient one for a number of instrument companies.

The coming decade, will most certainly show a multitude of new sensors and sensor principles, enabling the design of many new, useful products. These products will very often incorporate a large number of micro-electronic components, the demand for which will in turn lead to a growing and healthy electronics industry. Therefore, the greatest priority for the electronics industry must be to overcome the sensor lag. A look at the current world scene shows that interest and actual research with respect to the sensor field is very evident in Europe, but that the U.S.A. and especially Japan are making great strides to catch up in this field. Yet Europe, by making the right decisions with respect to the sensor field, could acquire a larger share of the future micro-electronics world market.

REFERENCES

(1) S. Middelhoek and D.J.W. Noorlag, Sensors and Actuators $\underline{2}$ (1981/82) 29.

(2) V. Zieren, D.J.W. Noorlag, S. Middelhoek and E. Wolsheimer, European Electronics $\underline{1}$ (5-1981) 10.

(3) Electronics $\underline{55}$ (jan. 13, 1982) 122.

(4) S. Middelhoek and D.J.W. Noorlag, J. Phys, E:Sci. Instrum., $\underline{14}$ (1981) 1343.

(5) K.E. Petersen, A Shartel and N.F. Raley, IEEE Trans. Electron Devices, ED-29 (1982) 23.

(6) P.W. Barth, IEEE Spectrum 18 (sept. 1981) 33.

(7) G.G. Guilbault, Y. Tomita and E.S. Kolesar Jr., Sensors and Actuators 2 (1981) 43.

(8) R.M. White, Proc. IEEE 58 (1970) 1238.

(9) H. Rahnamai and J.N. Zemel, Sensors and Actuators 2 (1981) 3.

(10) S.D. Senturia, S.L. Garverick and K. Togashi, Sensors and Actuators 2 (1981) 59.

(11) G. Heiland, Sensors and Actuators 2 (1982) ...

(12) E. Pettenpaul, et al., Siemens Res. and Dev. Rep. 11 (1982) 22.

(13) S. Middelhoek, D.J.W. Noorlag and G.K. Steenvoorden, Proc. Eur. Hybrid Micro-electronics Conf., Avignon (1981) p. 100.

(14) P.J. Severin, in Solid State Sensors, Kluwer, Deventer (1980) 49.

(15) S. Middelhoek and D.J.W. Noorlag, IEEE Spectrum 17 (feb. 1980) 42.

(16) V. Härtel: Opto-electronics, Mc.Graw-Hill, New York 1978.

(17) E. Heijne et al., Nucl. Inst. and Meth. 178 (1980) 331.

(18) J. Kemmer, Nucl. Inst. and Meth. 169 (1980) 499.

(19) D.J.W. Noorlag and S. Middelhoek, IEE J. Solid-State and Elec. Dev. 3 (1979) 75.

(20) O.M. Sprangers and D.J.W. Noorlag, Sensors and Actuators, to be published.

(21) G.D. Nieveldt, Sensors and Actuators, to be published.

(22) V. Zieren and S. Middelhoek, Sensors and Actuators $\underline{2}$ (1982) 251.

(23) E.C.M. Hermans, Sensors and Actuators, to be published.

Components, Devices and Subsystems Based on Surface Acoustic Wave Technology

Kjell A. Ingebrigtsen

ELAB, The Technical University of Norway, N-7034 TRONDHEIM-NTH, NORWAY

SUMMARY

The paper summarizes the most important applications of surface acoustic wave devices and describes the models and tools applied in their design and development.

1 INTRODUCTION

The first surface acoustic wave (SAW) devices were installed in operational systems 12 years ago. Since then a variety of devices in this technology has been developed and put into systems. With the exception of a few notable cases of high volume production, SAW-devices are mostly applied to rather specialized functions in equipment which is being manufactured in moderate quantities. A SAW-device is functionally similar to a custom designed analog integrated circuit. It is a specialized technology which requires skilled design and the cost required to develop a new device is therefore rather high. As demonstrated by Fig. 1 surface acoustic waves are well behaved. Design models and the manufacturing technology are accurate to 0.1 - 1.o per cent. Thus, the prospects of reducing the development cost are rather good. It will require more efficient CAD tools for automatic design, simulation, lay-out, and testing; and also that direct wafer writing is applied to the manufacturing in order to avoid mask expenses. SAW devices may then hit a much larger market of low volume, medium complexity equipment.

The most important contribution to the evolution of the SAW-technology, has been the development of single crystal lithiumniobate ($LiNbO_3$). The strong piezo-electricity and low acoustic losses of this material, and its excellent properties

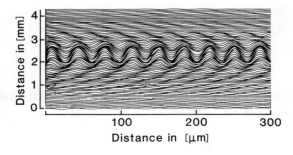

Figure 1.
Surface wave radiation
pattern at 100 MHz [1].

as a substrate for fine line metal patterns, have been of vital importance. From
a broad range of device concepts that have been demonstrated, a few remain as the
particularly significant. These are

- the implementation of transversal filters by interdigital SAW-transducers [2]
- the invention of the multistrip coupler [3]
- the development of the reflective array technology [4]
- the development of SAW-resonators [5, 6]
- the development of SAW-convolvers [7, 8, 9]

A substantial effort on acousto-electric devices such as travelling wave amplifiers, acousto-electric convolvers, memory correlators and related devices have so
far had little impact on practical systems. Mainly this has been due to complicated
manufacturing technology and poor reproducibility most often related to semiconductor material variations across the large interaction areas. However, there
is no doubt that some of these devices will be very useful if their fabrication
technology can be developed.

The present paper aims at describing the status of the SAW-technology. First some
important applications will be described, and next some of the tools used in the
design, manufacturing and testing of SAW-devices will be mentioned.

2   APPLICATIONS

The most important application of SAW-devices has so far been to signal filtering.
This includes frequency selective filtering whereby signals in a specific frequency
band are extracted and others are suppressed, and waveform matched filtering whereby
a specific waveform is extracted from other waveforms and noise. To the first cathegory belong most filters for communication systems. Quite often the task has been
to develop more competitive replacements to existing solutions. A typical example

is the IF-filter for home TV-receivers. To the second cathegory belong for example
pulse compression filters for radar receivers. SAW-devices are in this case often
the only solutions and their success are immediate even without pushing performance
and manufacturing costs.

Two different technologies are applied. Transversal filters are used for waveform matched filters and medium and large bandwidth frequency filters. Narrowband,
high selectivity filters are implemented by high-Q resonant circuits.

2.1 Transversal Filters

The principle of a transversal filter is shown schematically in Fig. 2. It is
a tapped delay line where the various taps are given various weights and added at
the output. Its impulse response is simply given by a finite train of impulses

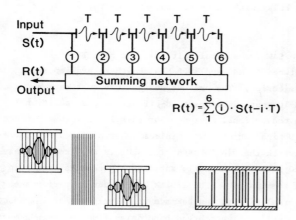

$$R(t) = \sum_1^6 \textcircled{i} \cdot S(t - i \cdot T)$$

Figure 2.
Transversal filter schematically. Weighting by apodization and selective withdrawal of strips, and multistrip couplers are often used.

each of amplitude proportional to the tap weight. Thus the duration of the impulse
response is given by the total delay through the filter. A transversal filter will
have multiple frequency responses separated by (1/T), the inverse time delay between
taps, unless the tap weights (i) are given a frequency dependence which suppresses
undesired responses. A transversal filter can be made phase linear. This is
required in systems for frequency- and phase modulated signals since deviations
from linear phase lead to intersymbol interference.

The interdigital surface wave transducer is an almost ideal solution to a transversal filter. The taps are metal fingers which detect and generate electric
potentials of surface acoustic waves travelling underneath on a piezoelectric substrate. The taps work most efficiently when the separation between the fingers is
one half acoustic wavelength. Signals from the various taps are summed directly

at the contact strips. Since an acoustic delay line consists of one input transducer and one output transducer, it is actually two cascaded transversal filters.
Various weighting techniques are being used. The two most common are shown in
Fig. 2. With apodization the metal strips overlap a fraction of the total beam
width proportional to the weight. The second method is the withdrawal weighting
whereby the effective number of strips in a tap is varied in accordance with the
weight. Since the beam profile transmitted from an apodized transducer is strongly
nonuniform, it is often necessary to insert a multistrip coupler between input and
output transducer. The beams coupled over to the adjacent track by the multistrip
coupler will always have a uniform beam profile.

Interdigital SAW-filters are most easily implemented with relative bandwidths
between 2 per cent and 30 per cent and with center frequencies between 10 MHz and
1 GHz. The acoustic wavelength which is about 3 μm at 1 GHz complicate high
frequency manufacturing due to lithographic line width resolution. At low
frequencies the SAW-devices are bulky with increased material costs, and other
solutions are available.

The most typical example of a SAW transversal filter is the home TV IF-filter.
It has a passband of 4.5 MHz between 34 MHz and 38.5 MHz and there are important
requirements to the stop band rejection. The present designs use substrate areas of
$\sim$ 5x10 mm, and they replace $\sim$ 8-10 coupled resonant LC-circuits and an additional
phase correcting network of the conventional design. For this large volume product
material costs are important. Designs which use a minimum substrate area are therefore important. In another approach the single crystal $LiNbO_3$ is replaced by glass
covered with a piezoelectric ZnO film. Technically satisfactory results have been
demonstrated with this material combination. The limits of SAW-device performance
are flexible. Wide band transversal filters can be made at the expense of increased
insertion loss. An example of a well designed broad band delay line has been demonstrated by Stocker et al. [10].

Impressive high frequency designs have been demonstrated by Urabe et al. [11] and
by Mitsuyu et al. [12]. Urabe et al. demonstrated well controlled designs at 0.9 GHz
and 1.9 GHz. Mitsuyu et al. have demonstrated that filters can be made at 4.4 GHz
using ZnO films on sapphire substrates.

Figure 3 is included to demonstrate designs which are pushing the design limits
of SAW transversal filters. Fig. 3a shows amplitude and group delay response of a
Nyquist filter intended for a professional TV-transmitter monitor [13]. The design
constraints are low amplitude and group delay ripple over a large relative bandwidth.
In the realization it was necessary to take specific precautions to suppress bulk
acoustic wave interference between input and output transducers.

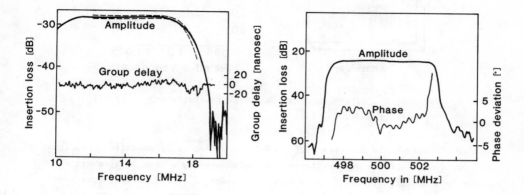

Fig. 3a Low frequency, broad band transversal filter on YZ-LiNbO$_3$. Transducers
designed with apodization and split
fingers.

Fig. 3b High frequency, narrow band filter
on YZ-LiNbO$_3$. Transducers thinned to 3/2
wavelength (input transducer) and 27/2
(output transducer) wavelength tap sepa-
rations. Output transducer 1600
wavelengths long.

Figure 3b shows the amplitude and phase response of a narrow band (1%) filter
realized on YZ LiNbO$_3$ [14]. The specific requirements were low amplitude and phase
ripple in the pass band, high skirt selectivity, and low insertion loss. The ultimate goal to the insertion loss made it desirable to implement the filter on LiNbO$_3$
rather than on quartz because of the lower propagation loss in that material. The
high skirt selectivity requires a filter response which is ~ 1600 wavelengths long.
The filter is designed with a long transducer which is thinned to a tap separation
of 13.5 wavelengths. The other transducer is designed to suppress the multiple
responses of the former. The problems encountered in the implementation were high
insertion loss, primarily due to bulk wave conversion, and low out of band rejection.
The latter was devoted to diffraction and undesired conversion to modes guided along
the contact strips. These are examples of secondary effects which may be encountered when the design rules are being stressed.

2.2 Resonator Filters

A surface acoustic wave resonator consists of two reflecting arrays as shown by Fig. 4.
By adding two transducers inside the resonator a two-port cavity resonator is
realized. A representative transmission loss which demonstrates the high selectivity of this structure is also shown. Since a single discontinuity at the surface

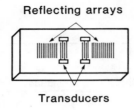

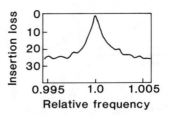

Figure 4.   Two-port reflective array surface wave resonator and a typical
two-port transmission loss.

leads to a considerable scattering of surface waves into bulk acoustic waves, it is
necessary that a surface wave reflector be implemented by a long array where each
element is a small reflector, typically of the order of 1%. The cumulative scattering from all elements of the array are phase matched only to the reflected
surface wave, and thus exhibit very small bulk wave losses.

Surface wave resonators with Q-values close to the material frictional loss limit
have been realized in this way. Unloaded Q-values exceeding 10^4 for frequencies
above 500 MHz are typical. It has been reported that these devices have been
developed to meet the strict specifications required for space operation [15].

Quartz is the most common material used for resonators. This is mainly caused
by the temperature variation of $LiNbO_3$ of 80 ppm/$^{\circ}$C which easily shifts the resonance
out of the original passband. The reflecting arrays may be realized by etching
grooves in the surface, by mass loading with strips of a different material, or by
combination of both methods by adding a different material into the grooves. This
latter technique provide a way of adjusting the response by making use of different
chemical etching rates or sputter etching rates of the substrates and the applied
strip material. On $LiNbO_3$ the reflectors may be implemented using metal strips where
the reflection is due to the shorting of the piezoelectric fields at the surface.
However, in this case the reported Q-values are nearly one order of magnitude poorer
than with grooved reflectors.

Resonators may be applied to frequency control of oscillators in much the same
way as in regular quartz crystal oscillators. The main advantage of the SAW
resonator appears to be its high fundamental frequency. Regular quartz oscillators
in the UHF range are overtone controlled, a solution which requires more complicated
electrical circuitry. Oscillators may be controlled also by inserting a SAW delay
line in the feedback loop of an amplifier. The delay causes a very fast phase
variation with frequency, and since the oscillator locks to a specific phase in the
feedback loop, a highly stable oscillator frequency results. This kind of oscil-

lator has been claimed to have the additional advantage of being able to deliver relatively high output power due to the much lower energy density inside the quartz crystal. So far the performance of SAW oscillators are about one order of magnitude poorer than comparable quartz crystal oscillators. However, considering the effort put into developing the art of manufacturing quartz crystal oscillators, SAW-oscillators have come a long way in a rather short time. Given the necessary effort they may well be competitive in the near future also for high performance applications. Figure 5 shows some results of aging studies of quartz delay line oscillators [16]. The lack of systematic drift is most disturbing.

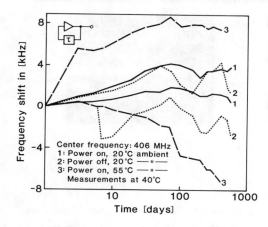

Figure 5.
Oscillator aging data.

2.3 Waveform Matched Filter

One of the corner stones of information theory is the concept of a matched filter. Assume that one is looking for a specific signal in a background of white noise. The matched filter gives the maximum improvement of the ratio of signal to noise power between filter output and input. The matched filter has an impulse response which is the time inverse of the signal waveform. The output is then the autocorrelation function of the waveform, and the processing gain is equal to the number of signal samples, that are coherently summed in the filter. Matched filtering has been used for a long time in radar and sonar systems. A short impulse which is required for range resolution, is encoded into a much longer waveform in order to reduce the peak transmitter power. The range resolution is recovered in the receiver by a matched filter which effectively compresses the long encoded waveform into a short pulse.

Surface wave devices are well fit to this purpose in radar systems, since they can accomodate radar bandwidths directly without the need of a buffering store. Processing gains exceeding 30 dB are available with SAW-devices. Figure 6 shows a

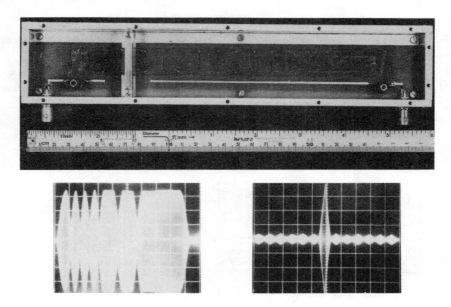

Figure 6.   13 bit Barker code SAW matched filter. Impulse response and
            matched filter response [17].

SAW matched filter for a 13 bit phase shift encoded Barker sequence. The impulse
response and the matched filter response are nearly ideal. It is worth noticing
the size of the quartz crystal required for the 52 μsec long impulse response.

2.4 Chirp Delay Devices and Subsystems

   Linear FM-sweeps are excellent for the encoding of radar pulses. The matched
filter response of an FM-sweep can be amplitude weighted in the receiver to give very
low time sidelobes. This restores the ability to distinguish between adjacent
targets of substanitally different scattering cross-section.

   The matched filter response of the FM-sweep is very sensitive to phase errors in
the receiver. The realization of FM-sweep expanders and compressors by discrete
components were therefore a headache for a decade untill SAW-devices were available.

   The SAW transducers with varying interdigital period is the most common way to
implement a linear FM chirp filter. Since the number of the periods in the FM-sweeps
often become very large, the transducer consists of many metal strips. Potential
problems with reflections, conversion into bulk waves etc. are avoided by using
slanted transducers whereby the various frequency components are spatially dispersed
on the substrate. This approach is applied to chirp filters with time bandwidth products (BT) up till 1000. Transducer chirps may be the only practical solution to fast

FM-sweeps. Stokes et al. have reported a 500MHz sweep in 250 nsec [18]. By applying careful fabrication technology Stocker et al. have demonstrated the design of slow sweep rate devices [19].

The most successful chirp devices have been made using reflective arrays [4]. Devices with time bandwidth products of 10^4 and otherwise near ideal performance have been ralized this way. The principle which is schematically shown in Fig. 7, is making use of frequency selective reflection from periodic arrays. To separate

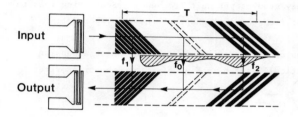

Figure 7.
Reflective array,
schematically.

the output signal from the input signal two arrays are applied each with 90° reflection. By increasing the array period with the distance from the transducers, the low frequencies are given longer delay than the high frequencies. This design has several advantages. Since the surface waves are travelling back to the output transducer, the delay for a given physical length is doubled . Since the various frequencies are spatially dispersed in between the two arrays it is possible to do frequency selective phase adjustment by adding extra delay in the path between the two arrays. Extra delay may be introduced by metallizing the surface to short out the piezoelectrid fields. This slows down the surface wave with about 1%.
Phase correction may be applied to slanted interdigital transducer in the same way since the various frequencies here also are spatially dispersed. In order to compensate for frequency dependent propagation loss and reflection loss, it is desirable to weight the reflection coefficient as funtion of position along the array. The first reflective arrays were using grooves etched into the surface. The reflection coefficient was varied by the use of ion beam etching to change the groove depth with position along the array. Other approaches are being pursued in order to develop fabrication techniques which do not require ion beam etching. Examples are withdrawal of strips combined with uniform sputter etching. This has two drawbacks, the scattering to bulk acoustic waves may be a problem, and when the effective number of strips per unit frequency is small, rather limited weight variations can be obtained.

In the reflective dot array, each groove or reflecting strip is replaced by dots. The dots may be metal dots, or they may be made by sputter etching holes in the substrate. In any case at least two mask steps will be necessary since the transducers require separate metallization. Encouraging results have been obtained by this approach. It is important to avoid coherent scattering from the two dimensional dot patterns into undesired directions, and noncoherent scattering into bulk acoustic waves may be a problem in some cases.

The reflective array approach is preferable for low sweep rates. Since the scattering from a groove or a dot partly is into a surface wave and partly into bulk waves, it is important to enchance the cumulative coherency of many scatterers in order to suppress the bulk wave scattering. This demands for many small scatterers operating coherently, or many grooves per unit frequency. Scattering loss to bulk waves depend also on the detailed physical shape of the groove or dot.

The most wellknown application of chirp devices is to radar pulse expansion and compression. This is schematically shown in Fig. 8. Since the transmitter

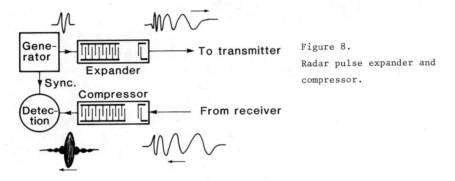

Figure 8.
Radar pulse expander and compressor.

is operated close to saturation the expanded pulse should be of uniform amplitude. The receiver matched filter must have the opposite chirp slope of the expander. However, often both the expander and the compressor devices are made down chirps (frequency decreasing with time) since this is easier to realize in SAW technology. The expanded pulse is then inverted before it modulates the transmitter. Time sidelobes in the compressed pulse are often reduced by weighting. Weighting can be carried out as a pure frequency filtering, but it is often incorporated in the compressor.

SAW-devices have made it possible to realize chirp transforms in an efficient way. The chirp transform is an algoritm which implements the Fourier transform . With SAW-devices it has the potential advantage of very fast real time transforms. It has a limited accuracy and a limited resolution. The principal configuration is shown in Fig. 9. The input signal is defined in the frequency and time domain by the bandwidth B and the time duration T over which the chirp transform is computed. The figure illustrates how a singular tone is transformed through the system and appears at the output in a time slot which is specific for the tone frequency. Since the frequency resolution is given by the integration time T, the number of

resolvable frequencies are BT, the time bandwidth product of the transform. The
swept LO must cover the bandwidth B over the time T. The convolution filter must
cover a frequency band of minimum 2B in the time slot 2T with the opposite chirp
slope of the LO. SAW-devices can be applied to generate the LO reference chirps
as well as to the convolution filter. A demonstration of a prototype system has
been given by Dolat et al. [21].

   With chirp transforms as building blocks it is possible to implement programmable correlators as shown in Fig. 10. This correlator will use at least 3 chirp

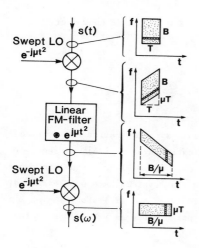

Figure 9.

Chirp transform implementation. The
diagrams to the right show the transformation of a signal defined in a
frequency range B over a time frame
T. The transformation of a pure tone
is also shown.

transforms. A rather impressive demonstration of a system of this kind has
recently been given by Estrick and Judd [22].

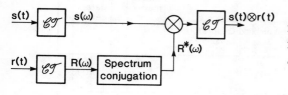

Figure 10.
Programmable correlator using
chirp transforms.

2.5 Convolvers and Their Applications

   The convolver shown schematically in Fig. 11 is a nonlinear acoustic device
which multiplies two counterpropagating surface acoustic waves. The central
electrode sums vectorially the product signals generated at all locations underneath it. The output signal which is modulated upon the second harmonic of the
input frequency is proportional with the convolution of the two signals except that
it is compressed by a factor of two. If one of the inputs is a reference signal,

the convolver performs a matched filtering of the signals at the other input.
Convolvers with integration times of 45 µsec and processing gains of 36 dB over
white noise have been demonstrated [23].   They have linear ranges which exceeds
the thermal noise floor by 70 dB.   This represents a tremendous equivalent com-
puting speed.   It corresponds to 4500 analog multiplications every 10 nanoseconds,
an instantaneous vectorial summation of the products, and presentation of the sum
with a dynamic range of 70 dB above thermal noise.

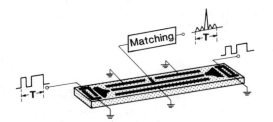

Figure 11.
SAW waveguide convolver.

Convolvers have been demonstrated for the code synchronization and message
aquisition in wideband spread spectrum communication systems [24].   A receiver
for DPSK-modulated signals need 2 convolvers, each one of length equal to two data
bit periods.   The performance of convolvers in test systems have been examined.
Figure 12a) shows a block diagram of a test receiver, and 12b) shows the
transmission test results of the system.   The performance is very close to the
theoretical predictions.

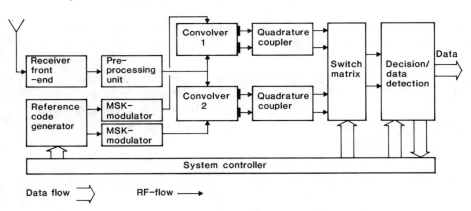

Figure 12a).   Spread spectrum DPSK demodulator using convolver for
matched filtering.

Characteristic for the applications shown above are high processing speed, moderate
and low processing complexity.   It is worth noticing that the speed of these SAW
processors are about one order of magnitude more powerful than projected future digital

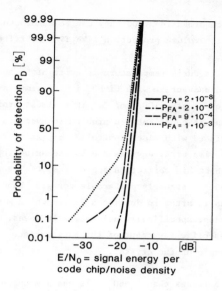

Figure 12b).
Test results demonstrating
processing gain close to the
theoretical limit using
1023 bit coding sequences.

processors based on Gigabit logic. Thus effort is being devoted to exploit the
potential advantage of SAW processors also in mainly digital systems [25].

Although the concepts of SAW-devices appear simple, it is worth being aware of
the auxiliary circuitry which are often required in a complete subsystem. This is
illustrated in Fig. 13 which shows the complete block diagram of a pulse expander.

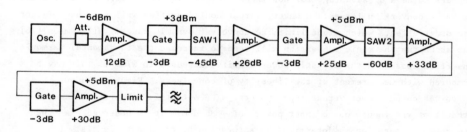

Figure 13. Block diagram of a SAW pulse expanded. Limitations to the
maximum input power combined with high insertion and expansion
losses require two cascaded SAW-devices.

3 PHYSICAL BASIS

3.1 Materials

The most important materials for surface wave devices are $LiNbO_3$ and Quartz. $LiNbO_3$
is strongly piezoelectric and it has very low acoustic propagation losses. Both
materials are relatively easy to handle and to apply submicron metal patterns to.

They are commercially available in ~ 25 cm dimensions and they can be manufactured
in wafers which is very convenient for high volume production like TV - IF filters.

$LiNbO_3$ is a ferroelectric material. Its Curie temperature is quite high, and it
is very stable and reliable . The main disadvantage with $LiNbO_3$ is its temperature
variation of the delay which is of the order of ~ 100 ppm/$^\circ$C. This almost prohibits
its use in frequency control circuits, and it constitutes a problem in hybrid systems where the analog circuitry is interfaced with clock controlled digital circuitry. A specific problem arises in reflective array devices due to the anisotropy of
the thermal expansion which leads to a shift in reflection angle from 90°. Because
of its high piezoelectric coupling, $LiNbO_3$ is strongly preferable for wide band
devices. A problem in these applications is often to develop a design which minimize the interference from other wave modes, specifically bulk acoustic waves.
Apart from the transducer pattern, and specific treatment of the back surface of the
crystal, this also varies with the cut angle of the crystal.

Quartz has somewhat higher propagation losses than $LiNbO_3$. It has a much weaker
piezoelectric coupling which may be an advantage in narrow band devices. The most
important property of Quartz is probably its temperature insensitivity. There are
cuts of quartz where the linear temperature coefficient of delay is zero. Quartz
is therefore the only practical material for resonators and oscillators.

SAW-devices may have linear dynamic ranges of more than 100 dB. At first this
appears to be a disadvantage for nonlinear devices like convolvers since it leads
to large conversion losses. However, larger nonlinearities will also have larger
3rd order effects which might limit the convolver dynamic range. Williamson [26]
has shown that harmonic conversion will tend to increase the fundamental harmonic
insertion loss when the acoustic power density exceeds 1-10 W/mm^2. This may be
compared with measurement of the linearity of convolvers. Figure 14 shows the
convolver output power versus input powers. The output is proportional to the
product of the inputs up to input powers of ~ 30 dBm. It is observed also that
the output saturation is determined by saturation of the input signals. By
estimating the delay line loss contributions, the 1 dB compression from linearity
occurs at an acoustic power of 21.5 dBm. This power is propagated through an acoustic waveguide with an effective acoustic beam width of ~ 40 μm. Since the penetration depth of the surface wave is ~ 1 wavelength or in this case 11 μm, the
maximum acoustic power density is 3.2 x 10^2 W/mm^2 which is two orders of magnitude
higher than Williamson has found. Propagation along a free surface is practically
dispersion free, and harmonic generation may take place coherently. In the acoustic
waveguide, harmonics would have different phase velocity and coherent conversion
is prevented. This should give a higher nonlinear threshold for the guided waves.

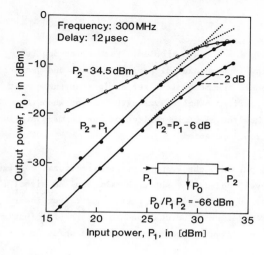

Figure 14.
Output linearity of a convolver
measured with long RF-bursts
of 5% duty cycle.

3.2 Device Modelling

The interdigital transducer (IDT) is the most important component in surface wave
device design. Figure 15 shows the 3 most common versions. Although a complete
transducer or transversal filter may be weighted by apodization, strip displacement,
strip weighting, etc., it will consist of elementary sections as shown in the
figure. The regular ID-structure has an electrical and mechanical period which is
one half acoustic wavelength at the center frequency. Each finger is an obstacle
in the acoustic path which will reflect surface waves. An infinitely long periodic
interdigital structure thus has a stop band at the center frequency. The width
of this stop band is determined by the piezoelectric strength and the mechanical
mass loading due to the metal fingers. In strongly piezoelectric materials such
as $LiNbO_3$, the electrode reflections are often a serious problem which prevents the
design of transducers with many strips and enforces solutions such as slanted

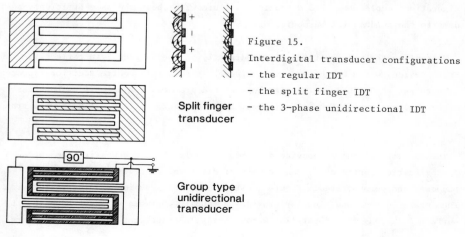

Figure 15.
Interdigital transducer configurations
- the regular IDT
- the split finger IDT
- the 3-phase unidirectional IDT

Split finger
transducer

Group type
unidirectional
transducer

geometries.

The split figure structure is a solution to this problem. Each electrode is divided into two strips thus resulting in a mechanical period which is one quarter of an acoustic wavelength. This cancels the mechanical reflections at the center frequency. The electrical period is still one half wavelength, except when the transducer is externally shorted. Thus the electric reflection can be made small by connecting the transducer into a low impedance.

Lately, substantial effort has been put into the group type unidirectional configuration. In this configuration a travelling wave is set up by the applied signal which causes a unidirectional excitation of the acoustic wave. Apart from lower losses the polyphase interdigital transducer may be designed to absorb efficiently the incident wave thus eliminating cumulative transducer reflections which otherwise give rise to tripple transit echoes.

The efficiency of a transducer depends on the spatial match between the electric fields from the surface wave and the electric fields from the interdigital structure. This may be expressed by a two dimensional overlap integral between the two field configurations. The theoretical calculation of the electric field from the interdigital structure is quite complicated. In nonregular structures, as transversal filters often may be, it is extremely complicated, and most often one has to rely on approximate methods.

The field configuration is very important for harmonic transducer operation. The potential from the interdigital structure has odd spatial harmonics which may be applied to harmonic operation of the transducer. The efficiency which is determined by the spatial harmonic components of the field is very sensitive to the detailed field distribution. For example, it turns out that the split finger configuration couples to the 3rd spatial harmonic with about the same efficiency as it does to the fundamental harmonic.

Most often the metal pattern is put directly on the surface. In reflective array devices which are using etched grooves it does not require additional processing to put the transducer pattern into etched grooves. This modifies both the mechanical reflection from the fingers, and the electric field set up by the transducer.

The multistrip coupler provide a method for multiplexing acoustic beams on a piezoelectric substrate [3]. A variety of different configurations have been proposed and demonstrated. Figure 16 shows 3 of the most important structures. The regular directional coupler provides a coupling between adjacent beams on the surface. It consists of parallel metal strips. An incident surface wave in one

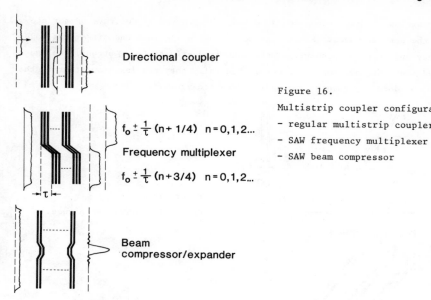

Directional coupler

$$f_o \pm \frac{1}{\tau}(n+1/4) \quad n=0,1,2...$$

Frequency multiplexer

$$f_o \pm \frac{1}{\tau}(n+3/4) \quad n=0,1,2...$$

Beam
compressor/expander

Figure 16.
Multistrip coupler configurations
- regular multistrip coupler
- SAW frequency multiplexer
- SAW beam compressor

track will give rise to an electric potential between the metal strips which in turn regenerate an acoustic wave in the neighbour track. It is usually modelled in terms of interference between normal modes. The beat wavelength between modes depend on the piezoelectric strength of the substrate. A 100 per cent coupler on LiNbO$_3$ requires a propagation length of ~ 45 wavelengths. Multistrip couplers on Quartz are impractical because of the weak piezoelectric coupling of this material.

The second structure shown in Fig. 16 is a frequency multiplexer. By offsetting one of the tracks to obtain a frequency varying phase difference at their inputs, it is possible to direct different frequencies into different tracks [27].

The last example shown in the figure is the multistrip beam compressor. Functionally it is similar to the regular directional coupler except that one track is narrow. In order to obtain complete transfer between the two different tracks, it is necessary to reduce the period in the narrow track and it is necessary to include diffraction in the modelling [28].

Reflective arrays are implemented either by grooves etched into the substrate, by metal dots, or by etched holes as shown in Fig.17. The latter techniques have been developed in order to be able to weight the reflection coefficient without depth weighting. In order to have control with second order effects, it is necessary to keep the reflection per groove or per strip low; typically less than 1-2 per cent. Unless there are many strips contributing, the reflection losses may be very high. In the design it is often necessary to include dispersion and non-synchroneous scattering loss. Dispersion occurs because the grooved or dot covered surface will retard the surface waves. Non synchroneous scattering into undesired wave modes is frequency and weight dependent. These undesired effects depend strongly on the detailed shape of the grooves or dots. Generally they are less important if the grooves and dots are smooth, and sharp edges are avoided.

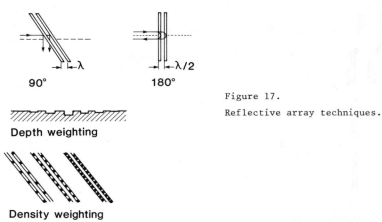

Depth weighting

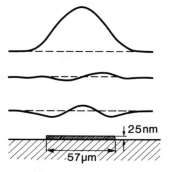

Density weighting

Figure 17.
Reflective array techniques.

The acoustic waveguide has become important in connection with convolvers. Since the efficiency of a convolver increases with the power density, it is advantageous to concentrate the acoustic power into a narrow strip. Since the convolver needs an electric output électrode, the combination with a metal strip waveguide is obvious. The trapping of the acoustic wave to the metal strip is caused partly by the piezo-electric effect and partly by the mechanical properties of the metal. By proper design it is possible to have these compensate and thus obtain a waveguide which is practically free of dispersion over a large fractional bandwidth [29].

Figure 18 shows schematically an aluminium waveguide on $LiNbO_3$ which is designed to have a minimum of dispersion for the fundamental waveguide mode at the center frequency of 250 MHz. Notice that this occurs for a very thin (25 nanometer) aluminium film. This guide may support 3 modes. Their amplitude profiles

Figure 18.
Metal strip waveguide with the
3 first propagating modes.

have been measured by a laser probe and are shown in the figure. The aluminium strip
adds some propagation losss to the surface wave. It has been measured to be
~ 0.3 dB/μsec at 300 MHz in this waveguide.

3.3 Design and Testing

The development of SAW-devices depend heavily on the use of computer aids.
Programmes for analysing the rather complex structures are necessary, and connections
between a specific pattern layout and its electrical response are also available.
The problem is often how to incorporate second order effects and when that could be
left to the judgement of the experienced designer. Different approaches are being
supported. An integrated system for design, measurement, and simulation have been
developed at Racal MESL [30].

Although the SAW-device requirements most often are given to its frequency
response, time response measurements have always been an important diagnostic tool
because most distrubances are caused by false echoes. However, the most accurate
measurements and the equipment most easily adapted to computer control is frequency
domain equipment like network analysers. Thus the preferred approach seems to be
frequency measurements followed by Fourier transform to obtain the time domain display, if that is necessary.

Figure 19a) shows a set up for time response measurements using a network
analyzer [31], and Fig. 19b) shows the result of a measurement which could hardly
been done in any other way[32]. It shows a plot of the output from a convolver

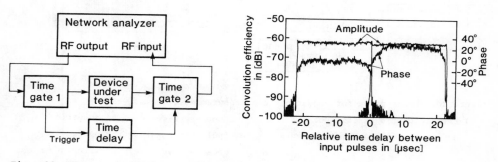

Figur 19. Frequency domain measurements of time domain responses.

  a) Measurement set up using a   b) Uniformity measurements of a
     network analyzer [31].   dual length convolver with
                                        separate outputs.

where the output electrode is divided in two strips of equal length with
separate outputs. Very often it is desirable to check whether the whole length of
the convolver is operating efficiently. This can be done by introducing a short

rf-impulse. By monitoring the amplitude and phase versus time of the output, it is possible to determine the spatial uniformity. The figure show amplitude and phase scans of both parts of the convolver. The difference in magnitude of the phase between the two parts is not significant, but the phase variation along each convolver length is. Each convolver output electrode is contacted through 12 taps. The location of these taps can be observed from the amplitude scans. It is observed that the convolver is quite uniform except for a phase shift towards each end of the output electrodes.

In research and development it is necessary to have a more detailed way of examination than purely electric terminal measurements. For this purpose laser probes [33] and scanning electron microscopes [34] have been fitted to do a point by point scan of the acoustic wave. The laser probe can be equipped to scan short pulses. A laser probe developed by H. Engan is shown in Fig. 20a). A 1GHz electrooptic modulator permit the use of very short acoustic impulses. Figure 20b)

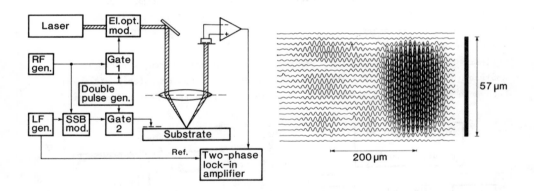

Figure 20. Laser probe for acoustic field measurements.

a) Principle of a laser probe for amplitude and phase measurements.

b) Laser probe scans of an acoustic impulse propagating through an acoustic waveguide. 3 different modes can be identified.

shows a scan of a short impulse propagating along an acoustic waveguide with propagation direction towards the right. The scan reveals that the incident signal has excited the 3 guide modes which due to dispersion are separated after 10 μsec propagation delay. The existence of the two higher modes will show up in the electric time domain measurements as two trailing echoes. Without the laser probe it would be very difficult to identify the origin of these echoes.

CONCLUSION

The purpose of this paper is to describe the performance of the most important surface acoustic wave devices in relation to their applications. The intention thereby has been to demonstrate that they are rather suitable solutions to specific applications.

Surface wave filters can be designed to satisfy strict specifications, and they have shown to be very reliable in operation. It is believed that the development of more effective computer aided design methods and manufacturing by direct wafer writing will lower the price on prototype development and small volume manufacturing.

Special devices like wideband chirp delay lines and convolvers have tremendous processing capacity which exceeds projected digital systems using Gigabit/sec logic by an order of magnitude. Such processing capacity is necessary to specific applications. It is conceivable to build complex systems based on such devices. This will require that effort be put into the interfacing with digital systems.

The available design models are quite good. Excellent agreement between simulations and measurements are obtained when the manufacturing is under careful control. The most common origin of device defects is connected to materials problems in quartz as well as in lithiumniobate.

It is to early to have a well formed opinion or the potentials of other materials/ material combinations like ZnO on Si, ZnO on Glass, ZnO on sapphire and ZnO on GaAs. If these combinations can be made uniform and stable, it will be possible to make quite interesting novel devices which combine digital with analog circuitry. ZnO applied to sapphire is interesting since it opens up a new frequency range for the SAW technology. With the high surface wave velocity of sapphire it is possible to reach the low microwave region 1–5 GHz by regular lithography. There is probably a market for an alternative filter technology in this frequency range.

REFERENCES

[1] Engan, H. private communication

[2] Smith, W.R. et al., IEEE Trans MTT-17, 856–864 (1969)

[3] Marshall, F.G. and Paige, E.G.S., Electronics Letters 7, 460–462 (1971)

[4] Williamson, R.C. and Smith, H.I., Electronics Letters 8, 401–402 (1972)

[5]   Ash, E.A., IEE. Symp. Microwave Theory and Techniques (1970).

[6]   Staples, E.K., Proc.Symp. on Frequency Control (1974)

[7]   Svaasand, L., Appl.Phys.Lett. 15, 300-302 (1969)

[8]   Cafarella, J. et al., Proc. IEEE 64, 756-759 (1974)

[9]   Defranoult, P. and Maerfeld, C., Ibid, 748-751

[10] Stocker, H.R. et al., Proc. 1980 IEEE Ultrasonics Symp. IEEE cat.no.
              80 CH 1602-2, pp 386-390.

[11] Urabe, S. et al., Ibid, pp 371-376

[12] Mitsuyu et al., Proc. 1981 IEEE Ultrasonics Symp. IEEE cat.no.
              81 CH 1689-9, pp 74-77.

[13] Rønnekleiv, A., unpublished (1976)

[14] Skeie, H., Unpublished (1974)

[15] Tanski, W.J. et al., Proc 1980 IEEE Ultrasonics Symp. IEEE cat.no.
              80 CH 1602-2, pp. 148-152.

[16] Rønnekleiv, A., unpublished (1982)

[17] Engan, H., private communication

[18] Stokes, R.B. et al., Proc 1981 IEEE Ultrasonics Symp. IEEE cat.no.
              81 CH 1689-9, pp.28-32.

[19] Stocker,H.R., et al, Ibid, pp. 78-82

[20] Solie, L.P., Proc. 1976 IEEE Ultrasonics Symp., pp. 309-312.

[21] Dolat V.S., et al., Proc 1978 IEEE Ultrasonics Symp. pp. 527-532.

[22] Estrick, V.H., and Judd, G.W., Microwave Journal 25, pp97-106, July 1982.

[23] Ingebrigtsen, K.A., unpublished (1981).

[24] Hjelmstad, J., Skaug, R., IEE Proc., 128, Pt. F, pp 370-378 1981

[25] Gautier, H., Proc. 1981 IEEE Ultrasonics Symp. IEEE cat.no.
81 CH 1689-9, pp. 206-219.

[26] Williamson, R.C., Proc. 1974 IEEE Ultrasonics Symp. IEEE cat. no.
CHO 896- ISU pp. 321-328.

[27] van de Vaart H., and Solie, L.P.,Proc 1975 Ultras.Symp.IEEE cat.no.
75 CHO 994-4SU, pp.322-326.

[28] Rønnekleiv, A., Engan, H., and Ingebrigtsen, K., Proc 1981 IEEE Ultrasonics
Symp. IEEE cat.no. 81 CH 1689-9, pp.305-310.

[29] Engan, H., Ingebrigtsen, K. and Rønnekleiv, A., Proc. 1980 IEEE
Ultrasonics Symp. IEEE ca.no. 80 CH 1602-2 pp. 77-81.

[30] Arthur, J.W., Microwaves, May 1982, p.143.

[31] Langecker, K., and Veith, R., Proc. 1980 Ultrasonics Symp. IEEE cat. no.
80 CH 1602-2, pp. 396-399.

[32] Grassl, H.P., private communication.

[33] Engan, H., IEEE Trans.Sonics & Ultrasonics, to be published.

[34] Veith, R. et al., Proc. 1980 Ultrasonics Symp. IEEE cat.no. 80
CH 1602-2, pp. 348-351.

X-Ray Lithography with Synchrotron Radiation

A. Heuberger, H. Betz

Institut für Festkörpertechnologie, Mikrostrukturtechnik; Lentzeallee 100,
D-1000 Berlin 33, Fed. Rep. of Germany

SUMMARY

   X-ray lithography at wavelengths between 0.5 and 5 nm is a simple one-to-one
shadow-projection process, with structural resolution as good as 0.1 μm under certain conditions. The main factors limiting the resolution are: Fresnel diffraction,
fast secondary electrons, the relatively-low mask contrast attainable in the soft x-
ray range, and - most important - the individual radiation characteristics of the X-
ray source in question. For a quantitive comparison of the different X-ray sources
which are applicable for lithography, numerical calculations of the generated resist
patterns are necessary. The simulation model on which the calculations are based has
to take into account the effects mentioned above, depending on the spectral distribution of the individual X-ray source, as well as the spectral absorption of windows,
mask substrates, and mask absorbers. Important boundary conditions in this connection derive from the present state of resist technology, especially in regard to
sensitivity, and from the necessity of compromising between source radiation power
and tolerable mask heating during exposure. Based on these considerations, a comparison between X-ray tubes, storage rings and plasma sources leads to the conclusion
that synchrotron radiation is superior to the others.

1 INTRODUCTION

   The process of X-ray lithography is a consistent and logical further development
of optical proximity lithography. This means that X-ray lithography involves simple

one-to-one shadow projection of an X-ray-transparent membrane with absorber structures onto a resist-coated wafer. Between mask and wafer is a small gap, the proximity
distance (typically 50 μm), to protect the mask against mechanical damage. The
typical arrangement appears in Fig. 1, which shows the similarity to optical proximity projection. However, there are two important physical facts which make X-ray
lithography much more difficult than the optical process:

(i) In the wavelength range of X-rays, the difference in transparency is not as
     high as in the visible wavelength region, which means that for the X-ray region
     at a given wavelength, there are no materials available which would be fully
     transparent in thicker dimensions (such as glass in the optical case), or which
     would fully absorb the radiation in very thin layers (such as chrome in the
     optical case).
(ii) There are no imaging optics available possessing a useful efficiency, which
     means that a condenser for homogeneous illumination of the wafer is not realiza-
     ble.

   These two points already summarize the general problems of X-ray lithography. As
a result of the first point, mask technology has to be modified. In order to obtain
a sufficiently-transparent mask substrate in this wavelength region, a light element
with low atomic number and low absorption has to be selected. Through use of a thin
foil covered with a relatively thick absorber structure, one can obtain the required
mask contrast.

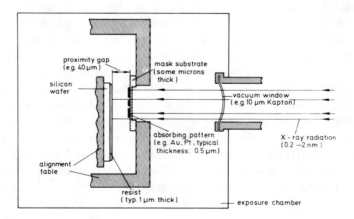

Fig. 1: Schematic exposure arrangement for X-ray lithography using synchrotron radia-
        tion

The wavelength region usable for the purpose of lithography is determined by the absorption properties of the mask substrate as well as those of the resist; this region corresponds to approx. 0.5-5 nm. For longer wavelengths, the limit is determined by the high absorption in the thin mask substrate; for shorter wavelengths, the limiting factor is the decreasing absorption in the resist layer, which becomes transparent in the case of harder radiation.

The second point mentioned above is of particular interest in respect to the subject of this presentation: The lack of useful optics such as lenses, in this wavelength range, means that the radiation has to be used in the same form (i.e., wavelength distribution and geometrical characteristics) as it is emitted from a given X-ray source. It is true that several research groups are investigating techniques for focusing soft X-rays, but the realization of a condenser with high efficiency and satisfying stability, applicable for production purposes, does not seem to be a solvable problem. Therefore, the successful application of X-ray lithography depends critically on the question of what kinds of X-ray sources are available, which is the main topic of this presentation.

2 RADIATION CHARACTERISTICS OF THE VARIOUS X-RAY SOURCES

For lithography purposes, X-ray tubes, synchrotrons or storage rings, and plasma sources are currently in use to greater or lesser degrees.

Most frequently used are X-ray tubes, since they involve a relatively well-known technique, and are comparatively inexpensive. However, in addition to several other disadvantages such as low intensity and difficult-to-change wavelength, they involve very unsatisfactory radiation characteristics. Specifically, they emit isotropic radiation, in combination with an extended area from which the radiation is emitted. This spot size of the source typically amounts to at least several mm. This leads to the well-known problems illustrated in the upper part of Fig. 2. The finite spot size denoted by 2r, acting together with the proximity distance s, causes blurring (a) and run-out (b). The term run-out refers to a systematical alteration in the lateral dimensions of the projected resist patterns. This effect is not as problematical as blurring, since it can be compensated for in principle by the mask layout, with a magnification scale which depends on the proximity distance as well as on the distance between source and mask. In addition, consideration has been given to the possible use of this effect for compensation of process-induced linear wafer distortions. But if there are uncontrolled variations of the proximity distance, and unflatness of the mask membrane, the resulting pattern distortions will enlarge very rapidly to values of 0.5 µm and greater. Clearly, a professional sub-µm technology is very hard to realize under such conditions. Attempts to improve the situation by increasing the distance between source and mask are very soon thwarted by the fact that the radiation density decreases quadratically with increasing distance.

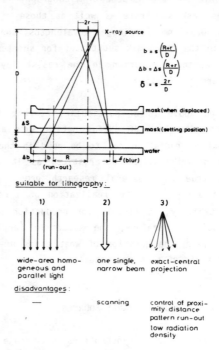

Fig. 2: General principle of radiation characteristics of X-ray sources

Therefore, it is necessary to consider the possibility of an alternative X-ray source offering better radiation characteristics. In the lower part of Fig. 2, three types of characteristics are shown which are more applicable in lithography: the best one would be the parallel and homogeneous floodlight of at least the size of the applied step-and-repeat field. As already said, however, there is only an extremely small chance of its realization.

The second case, the single beam with small aperture, is of the same quality in regard to the projection accuracy as the first case; here, however, there is the problem of a time-consuming and relatively-difficult scanning process. In addition, local mask heating is somewhat higher than in the case of homogeneous illumination, depending on the scan velocity.

The third case, exact central projection with extremely small spot size, solves the problem of blurring; but all the other problems of the X-ray tube are still present, especially, the exact control of proximity distance and mask flatness. Only in combination with a very high radiation density, in order to be able to enlarge the distance D between source and mask, would this case be suitable for sub-µm lithography. A fairly good approximation to this third case is represented by the

various kinds of plasma sources, which provide a very small focus spot and a high
intensity in comparison to X-ray tubes.

   Synchrotron radiation offers a mixture of the first two types of characteristics:
in the direction parallel to the orbit plane of the circulating electrons in a
storage ring, the radiation is similar to floodlight illumination, with a broadness
which is large compared to a silicon wafer. In the direction perpendicular to the
orbit of the electrons, the synchrotron radiation is collimated very narrowly. Fig.
3 illustrates the spectral power radiated from the storage ring DORIS in Hamburg, at
an electron energy of 1.5 GeV. In contrast to the monochromatic emission from an X-
ray tube, synchrotron radiation provides broad-band emission, similar to black-body
radiation. The maximum of the radiated power can be chosen freely by altering the
energy of the circulating electrons and the magnetic field in the bending magnets at
a given bending radius. The spectral intensity distribution can be calculated very
well, according to the classical formula of Schwinger and Sokolov /1/, which does
not need to be repeated here. The energy of 1.5 GeV provided by DORIS corresponds
with a radiation maximum of about 1 nm. For the most generally-used mask and resist
materials, this is the best choice, and ranges in the middle of the wavelengths
which are appropriate for lithography. Due to the higher magnetic field of the new
dedicated storage ring BESSY in Berlin (1.5 T), the electron energy of BESSY is only
800 MeV for the same wavelength distribution. In both cases, the so-called critical
wavelength, the best parameter to describe the spectral distribution independent of
the specifications of a particular storage ring, amounts to 2 nm.

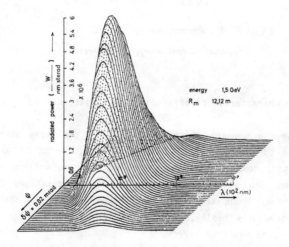

Fig. 3: Spectral and angular distribution of synchrotron radiation from the storage
       ring DORIS

The angle ψ in Fig. 3 represents the direction vertical to the orbit plane. The
spatial arrangement can be seen more clearly in the highly-simplified diagram of a

storage ring shown in Fig. 4. The distribution denoted by $\psi=0$ in Fig. 3 describes the radiation power emitted in the orbit plane. The radiation power decreases rapidly with decreasing ψ; one can see that the vertical collimation of the synchrotron radiation is better than 1 mrad. At a typical distance from the source point of 10 m, the soft X-ray component of the synchrotron radiation illuminates a horizontal stripe having a height of less than 1 cm. In addition, the radiation intensity is very inhomogeneous within this stripe in the vertical direction, as seen in Figs. 3 and 4.

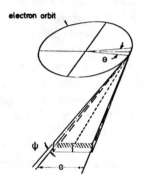

approximation of the angular spread $\langle \Psi \rangle$

for wavelengths near λ_c :

$$\langle \Psi \rangle \approx \frac{m_0 c^2}{E}$$

E = electron energy

$m_0 c^2$ = rest energy of an electron

Fig. 4: Emission characteristics of synchrotron radiation

Fig. 5 provides a summary of the properties of the X-ray sources under discussion. As opposed to the spectrum of a storage ring, the other sources are more-or-less monochromatic, depending on the target material as well as the gas within the discharge volume. The wide-band spectrum has some advantages in respect to the Fresnel diffraction; it is very important, however, that the spectrum of a storage ring can be changed very easily, e.g., to higher wavelengths for mask copying using thin e-beam-generated absorber layers.

The focus spot and the aperture have already been discussed; here, too, all advantages are on the side of the storage ring. In this case, the highest throughput is also achievable /2/. Only plasma sources, with their relatively small focus in combination with their especially high intensity /3/, and thus increased distance between source and mask, are also of interest for sub-µm pattern generation. With both interesting principles of plasma sources, laser-induced plasma and plasma focus, spot sizes down to 100 µm have already been realized /4/; adding the lateral instability by averaging over many pulses, the effective spot size should be lower than 1 mm. Therefore, the blurring effect is much lower than in the case of X-ray tubes. The overall pattern distortion will be discussed in sect. 5, referring to calculations of the resulting resist patterns. The intensities shown in the table of Fig. 5 indicate the power density utilizable in lithography; these are related to the source/mask distance shown above, and represent the maximum values realizable with currently-available equipment. The specifications of the plasma source have been chosen according to the plasma-focus system published this year by Physics International /5/.

Parameter	Storage ring	X-ray tube	Plasma source
spectrum	wideband typically 4–40 Å	monochromatic 8.3 Å (Al) 6.7 Å (Si) 4.4 Å (Pd)	wideband with monochromatic lines 7 Å, 12 Å
focus spot	0.5 mm	5 mm	≤ 1mm
aperture	1 mrad (with oscill. ≤ 5 mrad)	isotropic	isotropic
distance source/mask	10 m	30 cm	> 50 cm
blurring (prox. d. 50 µm)	< 0.01 µm	~ 0.5 µm	~0.1–0.2 µm
homogeneity	horiz: homogen. vert: Gaussian	homogeneous	homogeneous
pattern distortion	Fresnel	Fresnel + blurring	Fresnel (blurring)
intensity	10^{-1} W/cm²	10^{-3} W/cm²	10^{-2} W/cm²
time dependence	pulsed with high repetition rate (MHz) vert. scan : 1 Hz to 100 kHz	constant	pulses with 10^5–10^6 W/cm² (~20 ns, 0.1 – 1 Hz)

Fig. 5: Comparison of the various X-ray sources

Very important in connection with mask heating, treated in sect. 4, is the time dependence of the radiated power, as indicated in the last row of Fig. 5. The synchrotron radiation is pulsed because the circulating electrons assemble into bunches; the pulse frequency is determined by the number and circulation time of the bunches, and ranges up to several hundred MHz. In respect to lithography, however, synchrotron radiation can be assumed to be uniform in time, due to this high repetition rate. This assumption is not affected by the vertical scanning motion if the scan velocity is properly chosen (see sect. 3). However, the time dependence of plasma sources is very inhomogeneous, even in respect to lithography applications. All types of plasma sources provide short, very strong pulses having power densities up to 10^6 W/cm, with typical pulse lengths of 20 nsec. Compared to this, the repetition rate is very slow, amounting to about 0.1 - 1 Hz, depending on the expense of the equipment. This fact can cause severe problems in connection with mask heating (sect. 4).

3 METHODS FOR A WIDE-AREA HOMOGENEOUS ILLUMINATION WITH SYNCHROTRON RADIATION

There are several ways to achieve the same homogeneity in the vertical direction as is automatically obtained in the horizontal direction with synchrotron radiation:

(i)   Stimulation of vertical oscillations of the circulating electrons within the storage ring;

(ii)   Mechanical scan motion between the mask/wafer stage and the storage ring in the vertical direction;

(iii) Optical scanning with oscillation mirrors.

The stimulation of vertical oscillations, e.g. betatron oscillations, is a very effective and cost-saving method, but is limited in the achievable vertical height. This limitation results primarily from the design of the storage ring itself, e.g., the free gap of the bending magnets and instabilities; furthermore, the aperture angle should not be increased over 5 mrad for high-resolution purposes. Although our initial experiments in this direction at DORIS did not produce positive results, we are sure that this method will prove to be successfully applicable at BESSY, or for a compact storage ring. Our assumption is that step-and-repeat field sizes up to an edge length of 2 to 3 cm can be illuminated homogeneously in this way. This is significantly greater than the step-and-repeat fields of optical wafer steppers; however, we think that due to further progress in silicon technology, especially in the reduction of process temperatures, larger field sizes can be controlled even in the case that high resolution is desired. Therefore, an investigation of the alternative methods is necessary as well.

One of these methods, of course, is the mechanical scan motion of mask and wafer relative to the orbit plane of the storage ring. However, it is relatively difficult and expensive to move the mask and the wafer without disturbing the alignment accuracy. Only in the case of a compact storage ring does it seem possible that this could become the best and most economical method, since the whole storage ring itself could be moved as a unit.

Very interesting, even in the case of large step-and-repeat field sizes up to full-wafer exposure, are moving mirrors. Grobman /6/ is successfully using a curved mirror which deflects the beam downwards and collimates it in the forward direction. The curvature of the mirror increases the radiation intensity, especially in the case of small step-and-repeat fields, by focusing a broad horizontal stripe of synchrotron radiation onto the step-and-repeat field area.

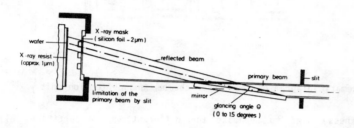

Fig. 6: Principle of arrangement using a plane mirror for vertical scanning

Our own investigations are based on plane mirrors, this being the simplest method to illuminate large areas homogeneously. Initial experiments were carried out at DORIS in Hamburg, using 3.3 GeV parasitically /7/. Fig. 6 shows a diagram of the beam path. To avoid rapid degradation of the mirror surface, the mirror chamber shown in Fig. 7 has been designed to meet ultra-high-vacuum requirements. The UHV chamber is therefore separated by two thin silicon windows, having a thickness of 2 µm, from the rest of the beam line. These windows have been produced by the well-known etching process for X-ray mask fabrication, e.g., as described in Ref. /8/. The mirror itself is driven by a piezo-mechanical actuator, which allows a reproducible positioning with an accuracy of better than 0.01 degrees. Additionally, the electronic control of the actuator permits regulation of the angular velocity profile of the mirror, which is necessary to achieve a homogeneous exposure. For these experiments, Zerodur mirrors of about 100 mm diameter, either sputtered with 50 nm gold or pure, were used.

Fig. 8 indicates the calculated power distribution for the described experimental arrangement, illustrating the spectral influence of a Zerodur mirror at different glancing angles, and also shows the spectral distribution of the emitted power (DORIS, 3.3 GeV, distance to source point: 27 m). The three silicon windows (each of

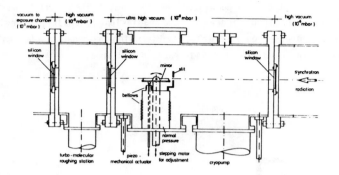

Fig. 7: Schematic diagram of the mirror chamber and deflection unit

2 µm thickness) used to separate the different vacuum partitions eliminate the wavelengths above 15 Å, and cause a sharp drop in the region around the silicon-K-absorption edge (6.7 Å). The wavelengths lower than this edge are absorbed very rapidly by the mirror with increasing glancing angles. The absorption of wavelengths between 6.7 and 15 Å remains very small within the whole range of relevant angles.

The most interesting parameter is the increase in the exposure time due to the intensity loss caused by the mirror. Fig. 9 illustrates this parameter, depending on the initial angle between mirror plane and storage-ring orbit, for various distances between mirror and mask, as well as for various mirror sizes. The numbers in brackets indicate the enhancement factors of the exposure time at minimum, compared to the direct exposure represented by the dotted line. The other parameters were, once again, 1.5 GeV using DORIS (critical wavelength: 0.19 nm), with a current of 50 mA, an overall beamline length of 27 m, and a step-and-repeat field size of 4x4 cm^2. The energy of 1.5 GeV (0.19 nm) is too high for lithography applications, and has been determined on the basis of parasitic use; the results, especially the ratio between the exposure times in the cases of a mirror and a direct beam, could possibly be improved if an energy suitable for lithography (1.5 GeV, λ_c = 2 nm) were to be selected. Our experiments will be continued at BESSY in Berlin as soon as the storage ring is in continuous operation. Summarizing the research done up to now, the results are very promising, and the increase in the exposure time is surprisingly low. In addition, there are some indications that the long-term stability of the mirror is high enough so that one can imagine an application in a production context.

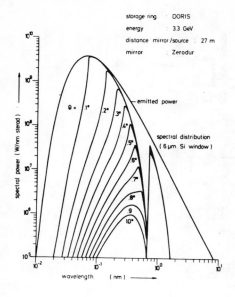

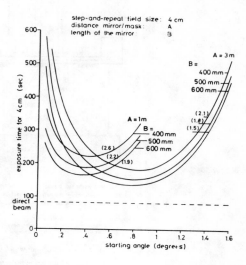

Fig. 8: Spectral influence of a
Zerodur mirror on a
synchrotron spectrum

Fig. 9: Enhancement of the exposure time
by using a plane mirror

4 MASK HEATING DURING EXPOSURE

A decisive problem, in connection with the application of alternative high-inten-
sity sources such as storage rings, is the question of whether there is an upper
limit to the radiation power because of lateral mask distortion due to mask heating
during exposure. Since it has not been possible up to now to measure the temperature
rise at high intensities - our beam line at DORIS in Hamburg does not provide enough
intensity, due to the length of nearly 30 m and currents below 100 mA - we have
developed a computer model to calculate the temperature rise under different conditions. Fig. 10 gives some illustrations of the physical effects which are taken into
account: absorption in mask membrane and absorber calculated for the realistic
spectrum, heat conduction through mask membrane and absorber, heat convection
through helium, and heat radiation from both sides of the membrane. The wafer chuck
and the mask rim are assumed to be at a constant temperature T_n, and to be connected
with a heat sink. The calculation procedure, as well as a detailed description of
the results, are to be published in the near future /9/; the most important results

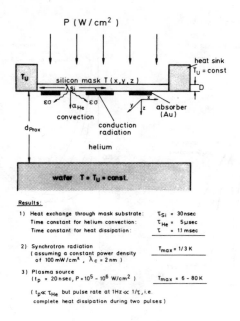

Fig. 10: Calculation of mask heating due to radiation absorption

for silicon masks are listed in the lower part of Fig. 10. The time constant for the heat exchange from one side of the 2-μm-thick mask membrane to the other, and from the absorber layer to the membrane, is very fast, only about 30 nsec. Therefore, temperature differences between the separated absorber areas can be neglected in realistic cases. The time constant for heat dissipation from the mask itself amounts to about 1 msec. The dominant contribution results from heat transfer from the mask membrane to the wafer, which is in contact with a heat sink via the helium gas. This leads, even in the case of very-intensive synchrotron radiation with a constant power density of 100 mW/cm^2, to a maximum temperature rise of the mask significantly below one degree.

It must be taken into consideration that the synchrotron radiation is not homogeneous in area, but is localized within the area of a small horizontal line. This source line must be scanned over the mask surface. Fig. 11 shows the resulting temperature distribution, depending on the scan velocity U in the vicinity of the source line, which is assumed to be at x=0. The numerical values in Fig. 10 are based on the case of a 2-μm-thick silicon membrane, helium cooling, and a proximity distance of 50 μm. If the scanning velocity is zero, the temperature distribution is symmetrical to the source line, and the temperature rise is very considerable. With increasing velocity, the distribution approaches the shape of a triangle, and the maximum temperature decreases. This decrease can be seen somewhat more clearly in

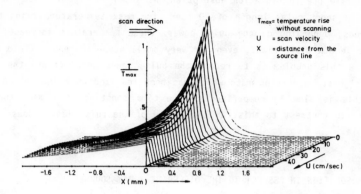

Fig. 11: Temperature profile in case of a line source moving with different velocities over a 2 µm silicon membrane

Fig. 12, where the maximum temperature $T_{max}(U)$ at a certain velocity, normalized to
the temperature without scan $T_{max}(0)$, is shown depending on the scan velocity U. The
conclusion of these considerations concerning synchrotron radiation is that, due to
the short time constants for heat dissipation, one must have relatively large scan
velocities of several 10 cm/sec in order to obtain low temperature rises. This seems
difficult to realize by mechanical scanning, which is a further argument for applying mirrors or betatron oscillations. However, this conclusion is only relevant for
extremely high intensities; in normal cases, difficulties are not to be expected
with heating of inorganic masks in connection with synchrotron radiation.

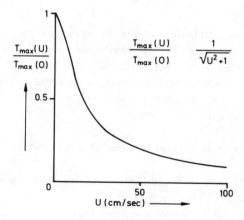

Fig. 12: Decrease of the maximum temperature with increasing scan velocity

In the case of plasma sources, however, the calculations indicate an unacceptable temperature rise due to radiation absorption, as shown in Fig. 10, item 3. Caused by the extremely short pulse length of 20 nsec, the mask temperature rises by about 6 to 60 degrees, depending on the pulse power. These temperature increases can cause intolerable length distortions, even for very small step-and-repeat field sizes. The way to solve this problem is to reduce the pulse power and increase the repetition rate; the highest repetition rate which has been realized up to now amounts to 1 Hz /3/, and this is slow in comparison to the time constant of 1 msec for the heat dissipation. In contrast to this, an increasing of the pulse length does not seem to be realizable.

5 PATTERN RESOLUTION IN RESPECT TO THE DIFFERENT X-RAY SOURCES

The importance of integrated-circuit process modeling increases rapidly with increasing packing density and circuit complexity. Future VLSI techniques require a complete and accurate quantitative description in order to optimize and control the technological process itself, as well as to take the best advantage of technology through circuit design. Therefore, for the development of new technological procedures - especially in the field of lithography, which is the most important process step in IC fabrication - the formulation of appropriate mathematical models is a necessity. Whereas, in the field of high-resolution optical lithography, a great deal of simulation work with important results has already been carried out, process modeling in the field of X-ray lithography is still in its very beginnings, and only a few papers have been presented up to now /10, 11, 12/. Therefore, for calculation of the structural resolution achievable with the various types of X-ray sources, we had to use the computer model developed by us which is described in detail in Ref. /13/.

The computer program now available is able to calculate the generated resist patterns for the three X-ray sources already considered, and for the most common X-ray resists such as PMMA, FBM, XXL15, etc., starting from the basic parameters such as type and thickness of absorber, mask membrane, window and resist, current and electron energy in the case of a storage ring, target material, primary power, source area in the case of an X-ray tube, etc.

The following effects can be calculated by means of a mathematical solution:

- spectral absorption of all materials involved, depending on the spectral characteristics of the source,
- Fresnel diffraction in respect to the complete spectrum,
- Fresnel diffraction including non-fully-opaque absorbers,
- extension of the focus spot of the source,
- range of the backscattered photoelectrons, using the simple depth-dose relation-

ship of Gruen,
- beveled absorber edges.

The relationship employed between dissolution rate and dose, for the most important X-ray resists, is based on empirical results.

To compare the high-resolution capabilities of the various X-ray sources, Figs.
13 - 15 show generated resist patterns for identical exposure conditions: 1-μm-thick
and 0.5-μm-wide gold absorbers on a 2-μm-thick silicon membrane, a proximity distance of 50 μm, and PMMA resist exposed with a dose of 1000 J/cm^3. The pattern
generated by an X-ray tube with an aluminum target is reproduced in Fig. 13. This
shows the geometrical shape of the different resist patterns, i.e., resist depth vs

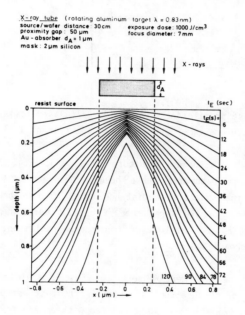

Fig. 13: Calculated resist profile in the case of an X-ray tube

distance from the middle of the absorber, due to the progress of the development
procedure at a fixed exposure dose. The parameter at the right side of each resist
profile refers to the development time. The resist has a thickness of 1 μm; zero in
the scale at the left side indicates the initial surface of the resist layer. It is
easy to see that there is no way to realize submicron patterns.

Using synchrotron radiation, however, conditions can be found for an accurate

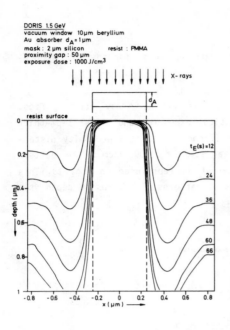

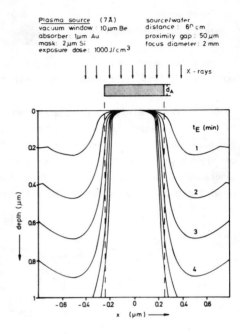

Fig. 14: Calculated resist profile in the case of a storage ring

Fig. 15: Calculated resist profile in the case of a plasma source

transfer of a 0.5-μm-wide absorber structure (Fig. 14). Fig. 15 represents the plasma source: the source parameters have been chosen according to the data of Physics International; the source/mask distance is 60 cm, in order to obtain the same exposure time as that of the X-ray tube (Fig. 13). Here, a pattern definition is obtained which is not as good as that of synchrotron radiation; however, it would be sufficient for submicron lithography.

These statements can be affirmed by some examples of achieved resist patterns, which show that the use of synchrotron radiation is the best method for carrying out X-ray lithography. These examples are not connected with an application in microelectronics, where only thin resist layers, mostly below 2 μm, are needed. The problem to solve here was to generate extremely thick resist patterns of several 100 μm, with lateral dimensions down to 1 μm, for applications in micromechanics. This work was carried out through cooperation between Siemens (ZFA), the Kernforschungszentrum Karlsruhe, and our Institute; a detailed description can be found in Ref. /14/. Fig. 16 shows needles of PMMA with a height of about 130 μm and diameters down to several microns. Fig. 17 illustrates the complete micromechanical structure, with a height of about 350 μm, which has been achieved up to now. Fig. 18 enlarges the area of Fig. 17 where the smallest dimensions (down to 1 μm) are located. The steepness of the walls is surprisingly good; the deviation from the vertical can be kept to below 1 μm over the total height.

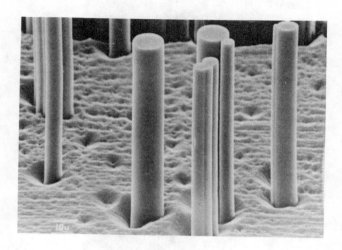

Fig. 16: PMMA needles with a height of approx. 130 μm

Fig. 17: Micromechanical structure with a height of approx. 350 μm

Fig. 18: Detail of the pattern shown in Fig. 17

6 THE COMPACT STORAGE RING

Synchrotron-radiation sources providing facilities for lithography experiments in Germany include the electron-storage ring DORIS in Hamburg and the electron synchrotron BONN II in Bonn. The 800 MeV storage ring BESSY, which is now nearly completed, will provide nearly-ideal conditions for lithography experiments. Outside of Germany, lithography activities are being conducted at several locations. To be mentioned first is the small storage ring at Brookhaven, where IBM is doing intensive research furthermore there are Orsay (ACO), Stanford (Spear), Dawesbury, and the photon factory in Japan. However, all these facilities are only well-suited for R+D activities; an application to circuit production does not seem to be practical.

Therefore, if X-ray lithography using synchrotron radiation is to have a chance for a broad application, it is necessary to realize special storage rings which are optimized in respect to costs and performance for lithography applications only. Currently-existing storage rings are specialized for purposes of fundamental research, with high demands on electron-beam focusing, time dependence, spectrum and beam stability, and so on. As opposed to this, a storage ring or synchrotron for lithography is a relatively simple machine, in which the beam size of the circulating electrons is artificially enlarged by stimulation of oscillations.

A subject for discussion is the size of a storage ring which is optimized for lithography. Here, a compromise has to be found between costs, applicable number of beam lines, and needed capacity. Many fundamental considerations in regard to circuit fabrication are involved in this matter, e.g., the question of whether it is acceptable for a fabrication line to depend on only one machine with its limited up-time, the problem of the best configuration in respect to a reasonably-clean room arrangement, etc.

In Germany the decision has been made, based on our early start in this direction, to realize a storage ring offering minimum size and costs. The basic principles for achievement of these conditions are the application of superconducting techniques as well as so-called weak focusing. We have started with the realization of the first prototype at BESSY this year; the first lithography experiments with this prototype will be carried out in 1985. The proposed specifications are listed in Fig. 19.

Parameter	Proposed value
λ_c	20 Å
current	300 mA
aperture (including betatron oscillation)	5 mrad
vertical movement	linear
lifetime	20 h
outer dimensions (including shielding)	2 m
number of beam lines	5 - 10
weight	< 10 t
power consumption	50 -100 kW
price without beam lines etc.	< 5 Mill. DM

Fig. 19: Specifications for the prototype of a compact storage ring

7 CONCLUSION

The considerations and facts presented lead to the following concluding statements:

- The most important advantages of X-ray lithography as opposed to optical-light processes, such as the exposure of very thick resist layers without the necessity of multilayer techniques, noncritical proximity distances (e.g. 100 μm), no problems with depth of focus, etc., can only be obtained using parallel X-ray light of high intensity, with the possibility of adjusting the source spectrum to the mask-and-resist combination. Synchrotron radiation offers the most advantageous preconditions by far for this purpose.

- With appropriately-designed storage rings, power densisties of more than 100 mW/cm^2 are achievable, which is about 2 orders of magnitude higher than in the case of an X-ray tube.

- The development of a compact storage ring, especially optimized for X-ray lithography use, is connected with the same range of necessary investment costs per exposure station as in the case of the X-ray tube.

- Plasma sources, extrapolated to the future, could represent the best compromise between performance and price; in this connection, however, there are still considerable development problems of a basic nature to be solved. Due to the huge pulse power, mask heating problems can arise.

- Assuming the availability of an optimal X-ray source, it is very possible to suppose that the process of X-ray lithography will represent the simplest and most economical approach for the fabrication of device structures in the sub-μm region.

- Technologically seen, the high resolution provided by X-ray lithography using synchrotron radiation can be controlled in overlay accuracy too, since automatic alignment systems with an accuracy of 0.05 μm (3 σ) are under development and seem very sucessful.

- This all does not mean that X-ray tubes would not be a good interim solution for a very early start with small-scale laboratory activities; in the future, though, meaning in the second half of the eighties, we think the compact storage ring will win the race.

REFERENCES

/ 1/ J. Schwinger, Phys. Rev. 75, Jun. 15, Nov. 12, and: A.A. Sokolov and I.M. Ternov, Synchrotron Radiation, Pergamon Press, 1968

/ 2/ W.D. Grobman, "Status of X-ray lithography", CH 1616-2/80/ 0000-0415, 1980 IEEE

/ 3/ H.I. Smith, Proc. Ninth Int. Conf. Electron and Ion Beam Science and Technology, St. Louis, Ed.: R. Bakish, The Electrochemical Society, May 1980

/ 4/ B. Wende, Physikalisch Technische Bundesanstalt, Berlin, private communication, and: G. Herziger, H. Krompholz, W. Schneider, and K. Schönbach, "A steady-state fluid model of the coaxial plasma gun", Physics Letters 71A, 54, 1979

/ 5/ Physics International, PIXI-10 pulsed plasma soft X-ray laboratory source; preliminary specifications, San Leandro, March 1982

/ 6/ W.D. Grobman, "An optical system for storage ring X-ray lithography", Electrochem. Soc. Spring Meeting, May 1982, Montreal

/ 7/ M. Bieber, H. Betz, and A. Heuberger, "Investigations of X-ray exposure using plane scanning mirrors", Intern. X-Ray and VUV-Synchrotron Radiation Instrumentation Conference, Hamburg, August 9-13, 1982

/ 8/ A. Heuberger, H. Betz, and S. Pongratz, "Present status and problems of X-ray lithography", in Festkörperprobleme XX, Ed.: Prof. D. J. Treusch, Vieweg, Wiesbaden, p.259, 1980

/ 9/ K. Heinrich, H. Betz, and A. Heuberger, "Heating and temperature-induced distortions of silicon X-ray masks", to be published in IEEE ED

/10/ P. Tischer and E. Hundt, "Profiles of structures in PMMA by X-ray lithography", Electron and Ion Beam Science and Technology, Eighth International Conference Proceedings, Ed.: R. Bakish, ECS, p. 444, 1978

/11/ A.R. Neureuther, "Simulation of X-ray resist line edge profiles", J. Vac. Sci. Technol. 15, 1004, 1978

/12/ J.R. Maldonado, J.M. Moran, "Resist Profiles in X-ray lithography", Proc. Microcircuit Engineering 80, Sept. 30, Oct. 1-2, 1980, p. 343, Ed.: R.P. Kramer, Delft University Press, 1981

/13/ K. Heinrich, H. Betz, A. Heuberger, and S. Pongratz, "Computer simulations of resist profiles in X-ray lithography", J. Vac. Sci. Technol. 19, 1254, 1981

/14/ E.W. Becker, H. Betz, W. Ehrfeld, W. Glashauser, A. Heuberger, H.J. Michel, D. Münchmeyer, S. Pongratz, and R. v. Siemens, "Production of separation nozzle systems for uranium enrichment by a combination of X-ray lithography and galvanoplastics", to be published in Naturwissenschaften

This work was assisted by the "Bundesministerium für Forschung und Technologie" of West Germany.

Silicon — the Perpetual Material for Micro-, Power- and Solar-Electronics

Diethard Huber and Werner Freiesleben

Wacker Chemitronic GmbH, Postbox 1140, D-8263 Burghausen, Federal Republic of Germany

SUMMARY

Silicon has become the unique raw material for today`s information processing techniques. The wide acceptance this semiconducting material has found in the electronics industry is based on its exceptional properties as well as on the abilities of the material manufacturers to improve their technologies according to material qualities, required to maintain appropriate progress on the learning curves of this industry. Today's capabilities and capacities of the materials manufacturers form a secure and reliable base to supply the material in large volume for the applications of today and tomorrow. Other semi-conducting materials will have difficulties to keep pace with silicon and to find their market niches.

1 INTRODUCTION

Worldwide production of defect-free, custom-doped, diamond-type silicon crystals has reached a rate of 350 carats per second or 2 200 metric tons per year with the average individual crystal weighing between 20 and 50 kg (or up to 250 000 carats). Technologies have been developed to control dopant (or "impurity") levels even to ranges of less than 0.1 parts per billion (1). We, therefore, deal with the most refined and purest material mankind has ever known.

Availability of such material presently enables and fuels production of roughly 18 billion $ worth of electronic devices which in turn are inevitable for manufacturing worldwide a volume of around 300 billion $ worth of products (computers, washing machines, watches etc., etc.) containing electronic controls or assistance in one form or the other. This volume represents 10 % of total world trade. Semiconductor materials other than silicon account for less than 5 % of the volume totally used. This paper discusses the general role of silicon in its major fields of application: power-, micro- and solar electronics as well as micromechanics.

2 SILICON AS SEMICONDUCTOR MATERIAL

Though Berzelius already in 1822 has isolated, named and described the chemical element silicon and though metallurgical silicon - especially as desoxidizing agent in the steel industry - was produced in millions of tons during the first decades of this century, it was not until the mid-forties when Western Electric (2) demonstrated completely new property aspects of the elemental material. By purification they prepared silicon with increased electrical resistivity and arrived at 30 - 50 milliohmcm. This material was used for rectifiers. Pearson and Bardeen (3) reached 100 ohmcm material in 1949 (Fig. 1) and were the first to describe silicon's semiconducting properties e.g. the activation energy of 1.115 eV at room temperature. This may be considered the birthday of our remarkable material. In the early 50's Siemens (4) developed and licensed technologies for the manufacturing of hyperpure polycrystalline silicon from purified trichlorosilane as well as the floatzone technique (5) for monocrystallizing the poly-silicon. They also prepared - by repeated floatzoning - materials with resistivities up to 150 kiloohmcm (6)

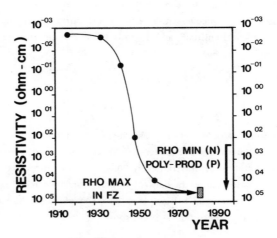

Figure 1. Resistivities ("impurity levels") achieved in
polycrystalline silicon during the past decades.

On this basis several big chemical companies started research and development for the manufacture of hyperpure semiconductor raw materials. Only few, however, persisted and generated - by technological contribution and accomplishments what today is called the "free silicon market". Fig. 2 demonstrates the progress achieved during the past twenty years: annual production of hyperpure polycrystalline silicon was worldwide increased from around 40 metric tons in 1962 roughly 100 times to near 4 000 metric tons this year. Within these two decades prices for poly-silicon dropped from around 500,-- \$/kg to near 50,-- \$/kg. Leading technologies elaborated by Wacker (metal bell-jars for deposition reactors) Dow Corning (deposition from dichlorosilane) Komatsu (deposition from silane) and Texas Instruments (gas phase deposition) led to todays capacities of total approx. 4 500 t worldwide and range from small units (several tons/y) to 1 800 t/y (Wacker). The ongoing task is to optimize quality and cost. Our poly production yields material with residual dopant levels typically far below the corresponding resistivities specified: 300 ohmcm for n-type and 3 000 ohmcm for p-type silicon.

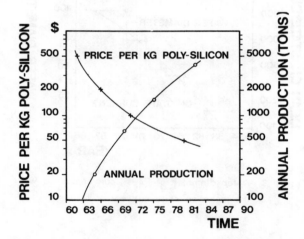

Figure 2. Price development and annual production
of polysilicon during the last 20 years.

3 POWER

Power electronics always demanded the purest, most homogeneous and most perfect material for application in thyristors, diodes, rectifiers etc. These requirements not only spurred well-controlled poly-silicon production at highest purities achievable but gave also rise to the development of larger diameters

of the crystals grown. Fig. 3 relates – during the past 30 years – the development of increasing silicon wafer diameters available and the "on state" current in thyristors (7). The power, thus handled could be increased by a factor
of more than 15 during the past 20 years with increasing wafer area available.
The parallel course of the two plots – by the way – has a simple physical
explanation: 1 cm^2 silicon of a silicon wafer can handle only a limited current.
Maximum allowable current depends on the silicon itself as well as cooling conditions and costs. From this plot a maximum allowable current of 25 - 30 amps
per cm^2 can be calculated. The figure also gives another interesting detail: it
reveals how long a time the power device industry needed to develop a new generation of devices and we can see that a 5" floatzone wafer will be required between 1983 and 1985.

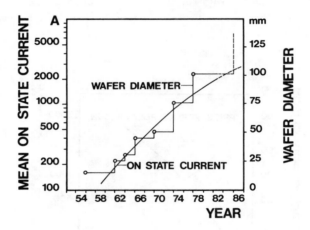

Figure 3. Wafer diameter of FZ-silicon compared with
the mean on state current of thyristors.

Sophisticated power devices need extremely homogeneous distribution of dopant.
To achieve this, transmutation doping (neutron-doping) was introduced in 1974.
This technique uses the nuclear conversion of Si isotopes (which are homogeneously distributed in silicon) into phosphorous dopant atoms by exposing undoped silicon crystals to a suitable flux of thermal neutrons in the core of a
nuclear reactor. The decay of unstable ^{31}Si into stable ^{31}P produces a material
with the required spreading resistance.

Yet another peculiarity of the power device industry is the sensitivity towards
carbon impurities. For power application, C-contents have to be lower than

5×10^{16} C-atoms/cm^3. Fig. 4 shows mean monthly values of our trichlorosilane production used for depositing poly-silicon. It is strictly monitored by a cumbersome analytical procedure. One receiving tank from trichlorosilane end column is sealed and sampled for a test deposition of a poly-rod. This poly-rod is floatzoned into a test crystal to get a sample where the carbon concentration can be determined accurately by infrared absorption. If the sample rests within our internal specification range (detection limit $< 5 \times 10^{15}$; upper internal limit 2.5×10^{16}; general specification: less than 5×10^{16} C-atoms/cm^3) trichlorosilane in said tank is okayed for further processing.

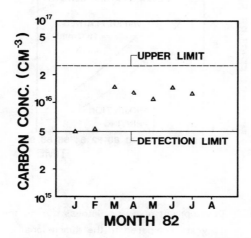

Figure 4. Carbon concentration (monthly average) of
todays large volume trichlorosilane production.

4 MICROELECTRONICS

Extremely tight controls of many wafer parameters govern the application of silicon in microelectronics – a field where silicon simply dominates. One example of concern is flatness. Shrinking device dimensions required more and more sophisticated mask equipment. Results of alignment and thus yields depend heavily on the availability of appropriately flat wafers. Fig. 5 shows the "learning curve" for wafer flatness and correlates it with the minimum device feature length during the past 12 years. Since different definitions for the wafer flatness are used in the industry, we want to emphasize that the values given here are absolute height differences measured as difference between the very top peak and the deepest valley on the wafer's landscape. Fig. 5 also indicates that sometimes during the end of the decade the 1 micron barrier might be broken. Another important aspect is the clean-

liness of the wafer surface. A 125 mm diameter wafer is free of minute particles under extreme control conditions. Clean room technology links the final stages of the
wafer production with the front end of device manufacturing.

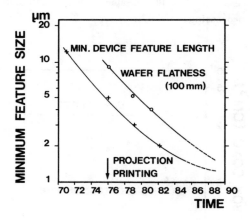

Figure 5. Development of wafer flatness during the last
12 years compared to the dimensional down-
scaling of device features.

There is no "standard quality" or "standard wafer" left. Silicon wafers have to
be material exceptionlessly falling within specification limits agreed upon between
user and producer. Device manufacturer and material producer have to learn from
each other how the best material should look like. Technology – or even custom –
tailored material becomes more and more important as the structures on the chips
become more complex. So for todays LSI and for the upcoming VLSI technologies
we developed not less than 6 crucible growing processes yielding different, controlled oxygen contents as well as different backsurface treatments for external
gettering purposes. To realise steady quality beyond the specification limits and
reproducible performance in the different device technologies stringent process controls are a must. Modern microelectronics is here a great help for the materials
manufacturer, so modern polishing machines or crystal pullers are unthinkable

without microprocessors or computer control. Fig. 6 shows as an example for the reproducibility of today's crucible growth processes the monthly average values for a material especially tailored for N-channel MOS application.

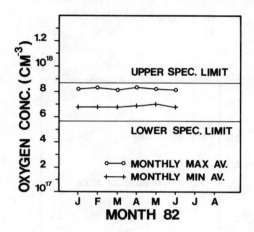

Figure 6. Oxygen range of N-channel MOS-material.

There is still another rather important quality parameter: wafer costs. The success story of microelectronics comprises manifold achievements and contributions: it's not only the design and build-up of devices or the multibillion $ contribution of the equipment manufacturers who engineered siliconwafer-specific machinery for the handling and the various processing steps to manufacture devices, or the accomplishments of the analysts and testing people, or the availability of suitable "electronic chemicals" - it's also the availability of appropriate polished silicon wafer area at reasonable costs which made all the astonishing results possible. Fig. 7 relates the prices for polished silicon area per cm^2 with the costs for finished IC area (with average values for memories). Of course, todays wafer (or silicon) manufacturers do not only produce just a few types of wafers: the product spectrum is extremely broad, indeed. There are the various forms and specifications for crucible and floatzone polycristalline silicon; there are many different ingot or crystal sales - whether diameters or electrical properties. There are all the various forms of wafers sold "as cut" or lapped; there are the hundreds of specifications for polished wafers having exact resistivities with tight tolerances in the extreme ranges from milliohmcm to kiloohmcm; n-type or p-type material of different crystal orientation, different diameters, thicknesses, surface conditions a.s.o. All the technology, the processes developed as well as the necessary logistics organized today are a prerequisit for a functioning supply but also a level of know-how and experience achieved for silicon which will make it extremely difficult - if at all possible - for any other substitute material to catch up or displace elemental hyperpure silicon as the dominating semiconductor material. And there are new, probably substantial volume applications just starting to emerge.

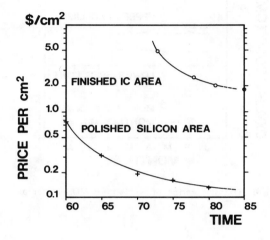

Figure 7. "Learning curve" of wafer manufacturing
compared to the curve for IC-manufacturing.

5 "SOLAR SILICON"

One of these new fields of application for hyperpure silicon will be photovoltaics. We have a roughly 100 Mio DM research program running to develop technologies for the production of solar grade silicon in forms which can be used to manufacture solar cells and photovoltaic systems. In table 1 we have linked up some of the major aspects for this use of solar energy: the upper scale measures cell efficiencies which can - theoretically - go up to 22 % for terrestrial uses. We differentiate and distinguish between various materials (and respective technologies to produce it and solar cells therefrom) and various applications according to the power supply desired. Basically we do not think, photovoltaics will solve mankind's energy problems - but we are convinced today that this form of using solar energy will have the potential for an interesting contribution. Let's look at the applications with a power demand in the range of milli-Watts (e.g. for calculators, watches etc.) - left

side of table 1. Here we could think of systems using amorphous silicon as power
generator. Characteristics of these systems will be: relatively short lifetime; p-n-
junction difficult to stabilize but cheap and very simple cell technology. There will
be an extensive market for such systems, however, with small "energy production"
in total number of megawatts generated. On the right side of table 1 are the
classical "space applications" with monocrystalline silicon or high-efficiency
thin films of other materials. In these applications efficiency or the <u>weight</u> of
a system and not the <u>costs</u> of the system is the dominant priority. In between is
a range from about 10 to a thousand kilowatts (even a few megawatts) where
flat panels can be used to supply the solar power. This is the classical range
for individual homes, isolated Diesel station replacements, irrigation jobs, water or power supply to villages with no link to public grids and the like.
These are the applications where we think photovoltaics will gain its merits.

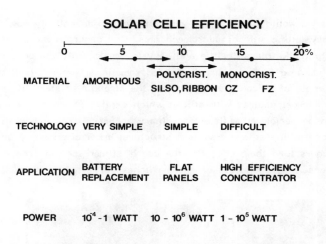

Table 1. Interrelation of factors governing the development
         of photovoltaics.

It will not be the several hundred megawatt power plants which will use photovoltaics but the millions of consumers of several hundred Watts or kilowatts.
This is the technological field we are concentrating our efforts on. Success to
provide appropriate systems will rest with the attainability of competitive sys-
tem-costs with conventional power supply from fossil fuels. This is also the
area, where we deem some form of crystalline silicon to be most promising. Of
course, the higher the power demand the higher will be the ratio of system
costs to material costs. For some hundred Watts this ratio could be 2.5; for

several hundred kilowatts it might go up to 12 - 15 with increasing importance of efficiencies.

Our research results, so far accomplished, hint optimism. Within few years we hope to be able to tell our engineers how to build a multimillion $-plant for the production of silicon material for solar cells which should - in appropriate systems - render a kilowatthour of energy for prices competitive to coalfired power plants. Aside from our cautious optimism it should be mentioned that our estimate of what can realistically be expected to be installed photovoltaic power towards the end of this century does not exceed several hundred megawatts worldwide.

6 MICROMECHANICS

Last not least we would like to draw your attention to an entirely new field of application for crystalline silicon: micromechanics. Our general conception regards silicon as something like glass: brittle and easy to break. Only the ones who saw a 45 kg crystal hanging from a 2 mm thick seed sometimes wonder about the enormous tensile strength of silicon - about 3 times the strength of steel. By etching we also achieve microstructures with silicon which exhibit an amazing dimension stability and can be performed with extraordinary accuracy. Yet - it wasn`t that obvious to use crystalline silicon also as a structural material and only until very recently researchers turn their efforts to the use of silicon for the preparation of X-ray masks, sensors, transducers or microswitches (8).

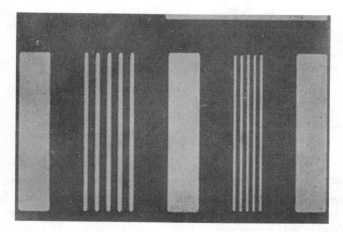

Figure 8. 1 micron linewith realized by X-ray lithography
on a silicon mask.

We think it's too early to discuss an "industrial" application already but applying the techniques learned in microelectronics – etching, diffusion, coating, metallizing etc. – to the manufacture of micro-parts with accuracies $< 1\ \mu$ apparently
opens an entirely new world for microengineers. The following slides shall illustrate just first examples of the new techniques and possibilities. We trust that the
electronic industry will rather rapidly engage in studies to exploit the unique properties of crystalline silicon in suitable combinations of microelectronics and micromechanics.

Silicon – purified to 99.999999 % and supplied as crystalline material – in our
judgement will even increasingly occupy an undisputed perpetual position in mankind's material spectrum and serve for probably centuries like the metals did
or earthenware, glass or organic materials like wood which not only provided
the mast for the Santa Maria sailing to new worlds but also the body for a
Stradivarius violin or the paper for this manuscript. We are spending up to
15 % of our sales in research and development to further increase our command
and understanding of the quality parameters of silicon in order to even better
service a most fascinating industry.

Acknowledgement: the authors have to thank Dr. A. Heuburger from the institute of solid state technology of the Fraunhofer Gesellschaft for the photos
on micromechanics.

REFERENCES

(1)   H. Silbernagel, Chem.Ing.-Technik, 50, (1978), 611.

(2)   J.H. Scaff, Met. Trans. of AIME, 1, (1970), 573.

(3)   G.L. Pearson and J. Bardeen, Phys. Rev., 75, (1949), 865.

(4)   F. Bischoff, U.S. Patent 3 146 123 (1961)

(5)   P.H. Keck and M.J.E. Golay, Phys. Rev., 89, (1953), 1297.

(6)   A. Hoffmann, K. Reuschel and H. Rupprecht, J. Phys. Chem. Solids,
      11, (1959), 284.

(7)   D. Eisele, J. Knobloch and K. Weimann, BBC-Nachrichten, (1981), 324.

(8)   K.E. Petersen, Proc. of the JEEE, 70, (1982) 420.

Thin Gate Oxide Technology

Eric DEMOULIN

Centre National d'Etudes des Télécommunications, CNET-GRENOBLE,
Chemin du Vieux Chêne, B.P. 42, F-38240 MEYLAN, FRANCE

SUMMARY

The trend towards scaled-down devices in MOS integrated circuits
pushes the development of ever thinner gate oxides. The characteristics
of thin (i.e. less than 40 nm) silicon dioxide are reviewed in view of
using such a dielectric in an industrial process ; attention is mainly
paid to reproducibility, control and reliability. The basic parameters
involved in gate oxide characterization are : thickness control, mobile
charges, fixed charges, interface traps and oxide traps. They are
discussed, as a function of processing parameters, for decreasing oxide
thicknesses.

1 INTRODUCTION

The evolution of the MOS technologies in the last decade has been essentially driven by the need for ever smaller devices. The well-known laws of scaling (1) tell us that the reduction in gate oxide thickness should follow upon reduction in channel length in order to obtain a well-behaved device. Substantial progress has been made regarding the industrial control of thin oxide technology (fig. 1). This leads to a crucial question for the evolution of the dielectrics : will the reliability issue, that was the main limiting factor in the past, allow the movement towards thinner gates to proceed in the eighties ? Or will severe limitations arise and preclude the use of very thin silicon dioxide as a gate insulator ? The goal of this paper is to review the problems connected with thin gate oxide technology.

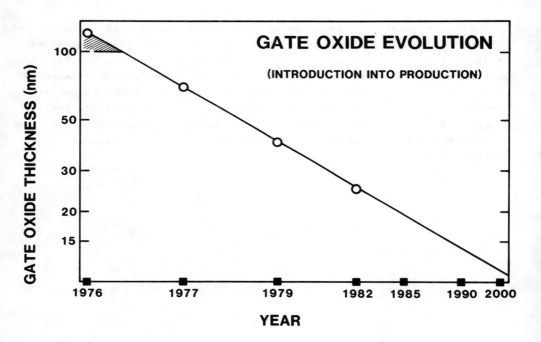

Fig. 1 - Evolution of the minimum thickness of gate oxide in mass-produced MOS integrated circuits.

Basically two distinct classes of devices have made a wide use of thin silicon dioxide layers. In the first, covering the MOSFET devices, the oxide acts as a dielectric and in normal operation no direct current flows through it. This has to be effective for any voltage within well-defined limits and at any time of the device operation, even under severe conditions (temperature, radiation in some cases, ...). To achieve this, a uniform and reproducible thickness must be controlled on an industrial basis, as well as defect density which must be improved upon. The related criterion is the distribution of breakdown electric fields. The threshold voltage of MOS devices provides the second basic tool for characterizing a given oxide. The thickness and the various charges in the structure (2), namely the fixed charges, the interface traps, the mobile charges and the oxide traps, must also be controlled on an industrial basis. To be more specific, the threshold voltage distribution should be as narrow as possible, while the initial values remain constant throughout the life of the device.

The second class of devices includes non-volatile memory devices (EPROM, EAROM, ...), in which current flows through the oxide to write or erase the information in the form of charges stored on a floating gate or in interface traps (MNOS devices). The current flow results from a tunneling mechanism, which only occurs in thin oxides (less than 4 nm for reasonably low applied bias), therefore called <u>tunnel</u> <u>oxides</u>. In these structures, the current flow itself generates secondary effects in the charge distribution, and premature degradation results, followed by accelerated threshold drift and catastrophic failure in some cases.

In any case, even if spectacular progress has been made in this field, the understanding of some basic mechanisms is still lacking. These will be mentioned throughout this review. Indeed, we are now facing a not-uncommon situation in the integrated circuits industry where technology has taken the lead over understanding.

Production of large scale circuits with thin gate oxides (25 nm) is under way, whereas very thin (tunnel) oxides can already be found in commercially available circuits. This means that high standards of quality, reproducibility and reliability have been reached, not only in a laboratory environment but also in the more demanding production environment. These are the criteria with which any new oxide technology has to be examined. This will be done in the following sections, where the basic characteristics of thin gate oxides will be reviewed as a function of processing parameters (fig. 2).

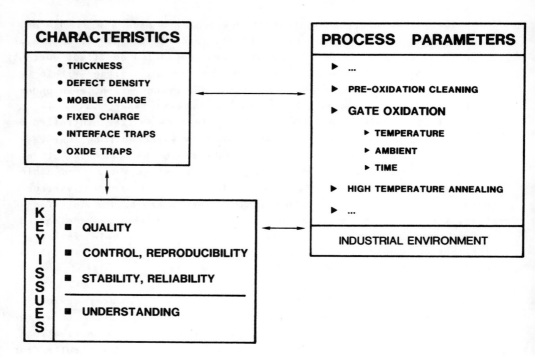

Fig. 2 - Criteria used for thin gate oxide evaluation.

It should be emphasized at this point that, from a technological point of view, it is definitely necessary to study the oxide characteristics in devices which have gone through the whole circuit manufacturing process. Some oxide properties heavily depend on process steps occurring either before or after the oxidation of silicon. These steps involve :

- bulk silicon characteristics (starting material, chemical impurities or structural defects introduced or removed by preceding high temperature treatments) ;

- surface characteristics (preoxidation cleaning, initial oxidation regime)

- bulk oxide and interface characteristics (generation or resorption of oxide and interface traps along with high temperature treatments, processes inducing radiation damage, ...).

The high sensitivity of MOS devices with respect to processing conditions makes it very difficult to compare results having different origins : no assurance can be obtained regarding the actual values of process parameters. For example, very small amounts of water can drastically modify the trapping characteristics of oxides (see section 7) ; unless very severe precautions are taken in controlling the ambient, it is not sure that unintentional humidity is not present in the oxidation chamber. This point probably explains contradictory results on oxidation behavior and oxide characteristics. Conclusions on the effects of some processing parameters should be drawn very carefully. Their applications in a different environment may lead in some cases to unexpected results.

Special attention must be paid to the gate material. Most results of advanced experiments are for aluminum gate devices. It has been widely observed that polysilicon gate devices, due to inherent differences in processes (like the higher temperature steps after gate deposition), often exhibit markedly different behavior. Since most of the coming MOS integrated circuits will be processed in a silicon-gate technology, we will mainly focus on this type of gate, with positive biases (the normal sign for NMOS and CMOS circuits).

2   OXIDATION KINETICS, OXIDE THICKNESS

Present gate oxides in the range of 10 to 100 nm are grown at temperatures between 950°C and 1100°C. The oxidant species are either dry oxygen or steam at atmospheric pressure, usually with the addition of a small amount of chlorine. In order to keep a reasonable control on the gate oxide thickness in a production environment, several conditions are to be fulfilled. Basically, the oxidation is designed in such a way as to minimize variations over the wafers in a batch and from batch to batch over long periods of time. The main point is to assure that the wafers are exposed to the same oxidizing ambient for the same time. To keep approximately constant the oxidation time while decreasing the oxide thickness leads to a reduction in the growth rate. This can be achieved either by using lower temperatures or by reducing the effective oxidant pressure.

Lowering the temperature is a main trend in VLSI processes, as it has the distinct advantage of limiting the spread of impurity profiles and the generation of crystallographic defects in the silicon substrate. Oxidation temperatures in the range of 800°C to 900°C have been reported (3-5). The question is whether the quality of the oxide or the oxide-silicon interface is affected or not.

Two methods were used to reduce the effective oxidant pressure. The
first one was to dilute oxygen (6-7) or steam (5) with an inert gas,
usually nitrogen or argon. Acceptable growth rates are obtained even at
temperatures as high as 1100°C, for final thicknesses in both insulator
and tunnel ranges (see fig. 3). Argon and nitrogen apparently give
comparable results (8). Suspicion exists against nitrogen in view of
possible reaction at the interface (9) ; on the other hand, argon offers
the advantage of being really inert, but its costliness presents a
definite drawback. The second method is based on the use of a low
pressure reactor (10). Good quality oxides were obtained in such a
system which is, however, more expensive than a conventional oxidation
furnace.

The methods described above allow process engineers to design an
oxidation process with reasonably good control and reproducibility as
far as thickness is concerned. Nevertheless, it is worth noting that the
basic mechanisms involved in the initial oxidation regime are still not
well understood. The classical linear-parabolic rate model (11) has been
confirmed experimentally for thick oxides. As far as thin oxides are

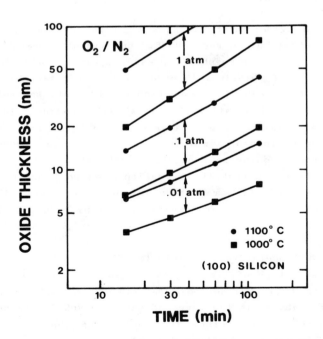

Fig. 3 - Oxide thickness versus time for silicon oxidation in diluted
oxygen. Data points are from reference (6).

concerned, not less than seven rate laws (4, 6, 7, 10, 12) have been proposed and more or less successfully used to describe the oxidation of silicon. In most cases the theoretical curve was fitted on scarce experimental points. Moreover, since measurement of the thickness of thin oxide layers is not a straightforward matter (13), the accuracy of these fits is not clear. Recently a systematic effort has been devoted to the acquisition of growth rate with an in-situ ellipsometer (14). These results confirm the enhancement in oxidation rate (fig. 4) which is usually observed with dry oxygen in the initial regime, typically for oxide thicknesses smaller than 20 nm (11). This initial regime is followed by a linear-parabolic one. The excess rate has been found to vary exponentially with oxide thickness (15-17), with a characteristic length in the order of 7 nm.

At the present time, no satisfactory explanation has been given regarding the physical mechanisms involved. Either a "parallel" oxidation process takes place in addition to the linear-parabolic process, or some of the basic mechanisms involved in this latter process (oxidant transport, surface reaction, ...) are enhanced for very thin

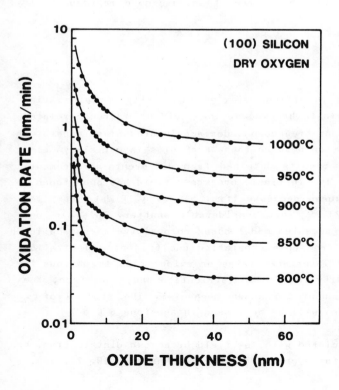

Fig. 4 - Oxidation rate of (100) silicon in dry oxygen. After reference (14).

oxides. Various suggestions have been made as to the origin of these enhancements :

- Does the charge distribution near the interface affect oxygen diffusion, if the diffusing species is charged oxygen (in either atomic or molecular form) ? If so, this would be consistent with experiments showing that oxidation can be accelerated or retarded by an adequate applied voltage (18). Although this may appear contradictory in the light of the observation that oxygen diffuses in a neutral form (19), it can be supposed that the molecules are converted to negative ions as they approach the interface.

- Are the structural characteristics near the interface responsable for these enhancements ? It has been suggested that an additional flux of oxidizing species may exist along either "micropores" (12) or "channel defects" (20).

Finally, it is worth mentioning the importance of the preoxidation cleaning procedure which has been shown to affect the surface and consequently the initial growth rate (21). Special attention should be paid to the cleaning procedure in order to assure good reproducibility on a day-to-day basis.

3 ELECTRICAL BREAKDOWN

Electrical breakdown in silicon dioxide is usually explained via three mechanisms : intrinsic breakdown, low-field or "defect"-related breakdown, and wear-out or breakdown under accelerated bias-temperature conditions. The distinction between these mechanims is not always clear (22) and comparison of results obtained from different measurement techniques is not easy. The intrinsic breakdown field has been found to increase as oxide thickness decrease (fig. 5), even if data compiled from various sources (23-25) show considerable scatter. This can be explained by the widely accepted model based on positive charge build-up in the oxide resulting from *hole generation due to impact ionization (22). The importance of interface states or oxide traps is obvious for the related charges modify the field distribution and, therefore, the current flow and subsequent breakdown mechanism. The problem of controlling these charges will be addressed in sections 6 and 7.

Defect density, associated with low field breakdown distribution, has been shown to increase for decreasing oxide thicknesses (10, 23, 24),

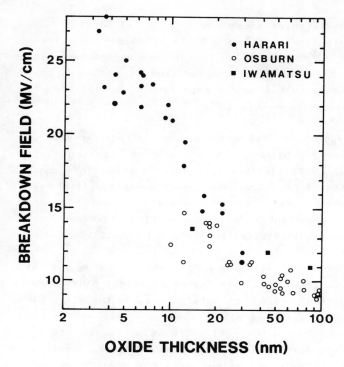

Fig. 5 - Breakdown
fields versus oxide
thickness. Data are
from Harari (24),
Osburn (23) and
Iwamatsu (25).

especially at less than 10 nm. The identification of those defects and
their origin are still a matter of speculation.

Heavy metal contamination is widely suspected to cause premature
breakdown in thin oxides. This is based on several observations : (i)
the better results obtained using FZ material with respect to CZ
material (26), (ii) the effectiveness of HCl flow in a double-walled
tube (27), (iii) the beneficial effect of "gettering" treatment with
back-side argon implant (28) or "sacrifice" oxidation prior to gate
oxide growth (26), (iv) the presence of metal-decorated stacking faults
at sites of breakdown occurence (29). Reports of these experiments
normally call for the obvious solutions : chase or remove the metallic
contaminants from the bulk, the surface and the environment of the
silicon wafers. A gettering technique, efficient cleaning and the use of
low-metal content products and materials are highly desirable for
reducing the density of defecte in the oxide.

The effect of oxide structure or ordering has also been invoked to
interfaces explain variations in breakdown phenomena. Lowering

oxidation temperature gives fewer breakdown defects (23,30) ; it is assumed that this results from a higher quality oxide due to a lower growth rate, related to the fact that low temperature oxides have higher density (31). This trend is, however, contradicted by at least one report (32), but this may result from the experimental method used for the defect detection (copper decoration).

It is also widely recognized that oxide charges affect breakdown statistics. Electron trapping in the oxide produces an opposing field which lowers the local high field. This decreases electron injection, thus reducing the likelihood of a breakdown. Water-related sites have been demonstrated to dominate oxide trapping characteristics (33) and are supposed to play an important role in controlling breakdown. Wet oxides have better breakdown statistics than dry oxides (12).

Since a very slight amount of water is sufficient to change the oxide trapping characteristics, it is likely that some excellent aluminum gate results are explained by the presence of water sites. On the other hand, breakdown results for devices having polysilicon gates are not as good as those obtained for aluminum gates (34). This is attributed to the elimination of oxide traps during the high temperature steps involved in a silicon-gate process. This is consistent with the poor breakdown statistics for oxides annealed in nitrogen at high temperature (35).

Finally, it should be pointed out that the beneficial role of oxide traps might possibly be exploited to good advantage : oxide traps could be selectively introduced into oxides in order to improve their breakdown characteristics (36).

4 MOBILE CHARGES

The problem of mobile ionic charges (mainly sodium and potassium) is not specific to very thin oxides. Since no HCl passivation (37) can take place at the oxidation temperatures currently envisaged for the future, the main effort, will have to involve improvements in materials and technologies to avoid sodium contamination in the oxidizing environment.

New difficulties may very well arise from the use of new gate materials, such as refractory metals. The phosphorus diffusion from N+ polysilicon into the upper layer of the gate oxide will no longer assure the gettering effect usually attributed to it. Consequently, some passivation could be obtained using a two-step technique (32) with oxide growth in a HCl ambient followed by high-temperature treatment to activate the passivating role of chlorine. However, this high temperature step will complicate the design of a VLSI "superficial" process.

5 FIXED CHARGES

The measurement of fixed interface charge density (Nf) becomes very difficult for thin oxides. For thick oxides, it is usually determined from the measured flat-band voltage (V_{FB}) through the relation :

$$V_{FB} = \phi_{MS} - qN_f/Cox \quad (I)$$

where ϕ_{MS} is the work function difference between the gate and the bulk and Cox is the oxide capacitance per unit area (= $\epsilon ox/Xox$; ϵox is the oxide permittivity and Xox the thickness). For a 20 nm oxide with a Nf value of 5 X 10^{10} cm^{-2}, the last term of equation (I) is equal to 46 mV. To determine Nf within a 10 % accuracy, ϕ_{MS} must be known with a less than 5 mV error. Unfortunately, the experimental values for ϕ_{MS} are widely scattered, and have been shown to depend on processing conditions (38), with apparent variations exceeding 500 mV. This spread explains part of the inconsistencies of reported results and particularly the occurrence of "negative" fixed charges.

Conversely, the large value of Cox and the resulting small value of qNf/Cox makes this term less and less critical for threshold voltage control of thin oxides. Those points justify the low level of interest for fixed charges.

Roughly speaking, the fixed charge density follows the familiar Nf-temperature triangle (2), just as thick oxides do. However it should be noted that slightly higher values are obtained after oxidation and fast pull-out in the oxidizing ambient (39) ; the excess value of Qf increases for decreasing oxide thicknesses. The effect of post-oxidation annealing in nitrogen is similar for both thick and thin oxides, with an initial Qf decrease followed by an increase for longer times. This upturn is not present with argon. Typical results of 1 to 2 X 10^{11} charges/cm^2 have been measured on about 30 nm oxides (39).

No clear theory has been found to explain the observed trend. The higher values of Qf for thin oxides are likely to be related to the growth mechanism associated with the fast initial oxidation regime.

6 INTERFACE TRAPS

Interface traps (Nit) do not exhibit significant behavior differences for thick and thin oxides, at least regarding their structural type. Some increase in Nit takes places for thinner oxides, but this still remains at a rather low level (39). The beneficial effect of annealing in forming gas is not altered when the oxide thickness is reduced.

The main problem with interface traps results, under given conditions, from interface trap generation associated with current flow and charge trapping in the oxide. This will be treated in the following sections.

7 OXIDE TRAPS

Electron trapping in the oxide presents both drawbacks and advantages. On the one hand, it induces threshold voltage shift due to the build up of a negative charge in the oxide bulk. Prolonged electron injection results in a "turn-around effect" (40, 41), i.e. a decrease in threshold voltage, which has been attributed to the formation of positively charged interface states. On the other hand as discussed in section 2, the presence of negative charges in the oxide tends to lower the local electric field and consequently to increase the breakdown voltage. However, it should be noted that this is a time-dependent property, which only appears after significant electron injection, i.e. after considerable stress.

The intrinsic electron traps in silicon dioxide have been shown to be due to water-related centers (33, 41, 42). It must be kept in mind that even a small amount of water is sufficient to seriously affect the oxide properties (42). This may arise unintentionally during oxidation : some water vapor may be present in a supposedly dry gas or may diffuse towards the wafer through the walls of the quartz tube (43) or by backstreaming from the open tube end. This may justify, in an industrial environment, deliberate use of an oxidizing ambient which contains a minimum amount of water so that the introduction of water sites in the oxide can be accurately controlled.

The majority of the studies on oxide trapping has been to aluminum gate structures. For such structures, the flatband voltage shift has shown to be proportional to the square of the oxide thickness (41) which is a very favorable situation for thin gate oxides. Very dry oxides exhibit very little flatband voltage shift and no turn-around effect (42). Post-oxidation heat treatments reduce oxide trapping, presumably by driving out water-related centers.

The same behavior is observed in the case of polysilicon gates. Compared to an aluminum gate process, a silicon gate one always includes several high temperature steps which tend to lower the oxide trap density and bring about an absence of the turn-around effect (41). Shifts in flatband voltage have been shown to be a logarithmic function of oxide thickness (34) and consequently very small for thin gate oxides.

In conlusion, electron trapping will not adversely affect the performance of MOS devices made using very thin oxides, since the total number of traps is low for this case. Nevertheless, even when dealing with thin oxides, the density of water-related traps must be controlled all along the process. This is true not only during the high temperature steps, but also when the wafers are cooling down to room temperature (44).

The low density of electron traps in silicon gate MOS devices does lead, however, to a relative degradation of the breakdown statistics (see section 3). Finally, it should mentioned that after injection of hot electrons, oxides become slightly leaky (45), especially very thin ones (5 nm).

8 RADIATION EFFECT

Classically radiation damage is a concern when dealing with devices being exposed to the severe environment of ionizing radiation. Recent developments in processing technologies have introduced new sources of radiation : electron beam lithography machines, X-ray aligners, reactive ion etchers, sputtering systems, ... all of which directly irradiate the wafers. The sensitivity of interfaces and thin oxides to these radiation-inducing technologies must be carefully evaluated.

Very small threshold voltage shifts are expected, after radiation, in MOS devices with very thin oxides. These were reported to be proportional to either Xox^2 (46) or Xox^3 (47). The radiation effects are generally attributed to hole trapping closed to the silicon-silicon dioxide interface (48). On this basis, hole trapping in silicon dioxide has been used to characterize the sensitivity of the structure with respect to radiation damage. However it has to be pointed out that, even if such an experiment shows significant hole trapping, it should not be taken for granted that large threshold voltage shift would occur in a radiation-rich environment. Actually, this will only happen if two phenomena are present simultaneously. On one hand, holes must be generated in the oxide by the ionizing radiations ; but it should be noted that decreasing the oxide thickness decreases the volume in which hole generation takes place. On the other hand, holes have to be trapped in the traps present near the interface. This trap density sould be kept as low as possible, for example by adopting low temperature processes (48).

Electron-beam lithography and reactive ion etching, which are potential candidates to fabricate micron and submicron devices, use energetic particles which, directly or by generated X-rays, ionize material as they pass through it. Electron irradiation induces a large threshold voltage shift (49) which is not completely removed by a 400°C final anneal. Moreover, if most of the charged defects are annealed out, it is worth mentionning that induced structural damage is still present, as well as neutral electron traps (50). These traps, if charged during the operation of the device, lead to threshold shift and reliability problems.

Reactive ion etching (RIE) has been shown to produce positively charged traps at the silicon-silicon dioxide interface and neutral traps throughout the bulk of the oxide (51). These traps can be annealed at 600°C. Moreover it has been reported (52) that RIE induced traps are generated by photons that are absorbed in polysilicon. Therefore, when RIE is used to delineate the polysilicon that is covering the gate oxide, this silicon layer properly shields the oxide from radiation damage.

9 CONCLUSIONS

The characteristics of thin oxides for MOS silicon gate devices habe been reviewed. Oxide thickness, breakdown statistics and charges in the oxide bulk and at the interface are the parameters to be controlled and reproduced. Contradictions may appear in results coming from various sources. Nevertheless, the most probable effect of the main processing steps are synthetized in figure 6. A few guidelines can be deduced from it to design a thin oxidation process.

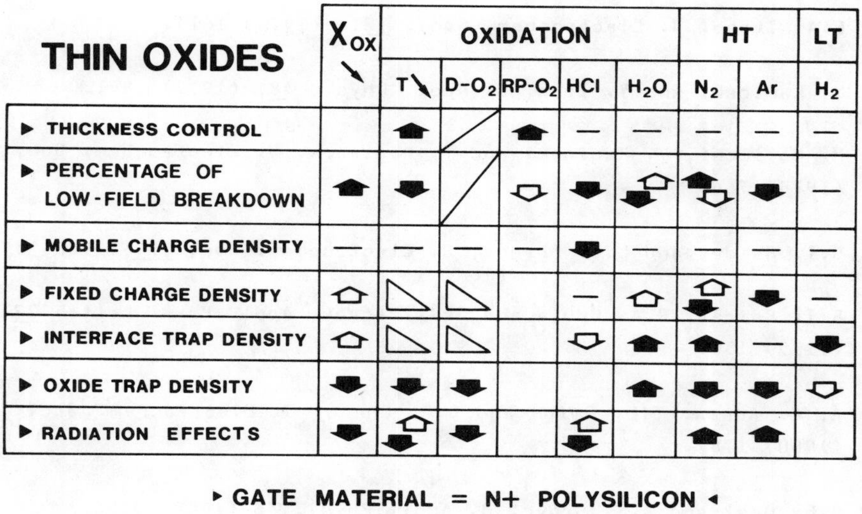

THIN OXIDES	X_{OX}	OXIDATION					HT		LT
		T	D-O₂	RP-O₂	HCl	H₂O	N₂	Ar	H₂
▶ THICKNESS CONTROL		⬆	/	⬆	—	—	—	—	—
▶ PERCENTAGE OF LOW-FIELD BREAKDOWN	⬆	⬇	/	⇩	⬇	⇧⬇	⬆⇩	⬇	
▶ MOBILE CHARGE DENSITY	—	—	—	—	—	⬇	—	—	—
▶ FIXED CHARGE DENSITY	⇧	◺	◺		—	⇧	⇧⬇	⬇	
▶ INTERFACE TRAP DENSITY	⇧	◺	◺		⇩	⬆	⬆	—	⬇
▶ OXIDE TRAP DENSITY	⬇	⬇	⬇			⬆	⬇	⬇	⇩
▶ RADIATION EFFECTS	⬇	⇧⬇	⬇		⇧⬇		⬆	⬆	

▶ GATE MATERIAL = N+ POLYSILICON ◀

Fig. 6 – Effect of processing on thin oxides characteristics.

REFERENCES

(1) R.H. Dennard, F.H. Gaensslen, H.N. Yu, V.L. Rideout, E. Bassous and
     A.R. Le Blanc, IEEE J. Solid-St. Circ., SC-9 (1974) 256.

(2) B.E. Deal, J. Electrochem, Soc., 121 (1974) 198 C.

(3) Y.J. Van der Meulen, J. Electrochem. Soc., 119 (1972) 531.

(4) M.A. Hopper, R.A. Clarke and L. Young, J. Electrochem. Soc., 122
     (1975) 1216.

(5) E.A. Irene, J. Electrochem, Soc., 121, (1974) 1613.

(6) Y. Kamigaki and Y. Itoh, J. Appl. Phys., 48, (1977) 2891.

(7) M. Horiuchi, Y. Kamigaki and T. Hagiwara, J. Electrochem. Soc, 125
     (1978) 766.

(8) S.I. Raider and L.E. Forget, J. Electrochem. Soc, 127 (1980) 1783.

(9) S.I. Raider, R.A. Gdula and J.R. Petrak, Appl. Phys. Lett., 27
     (1975) 150.

(10) A. C. Adams, T.E. Smith and C.C. Chang, J. Electrochem. Soc, 127
     (1980) 1787.

(11) B.E. Deal and A.S. Grove, J. Appl. Phys., 36 (1965) 3770

(12) E.A. Irene, J. Electrochem. Soc, 125 (1978) 1708.

(13) P. Ged, A. Vareille and M.H. Debroux, ESSDERC 82, München,
     Germany (1982).

(14) The date were obtained by H.Z. Massoud and J.D. Plummer (Stanford
     University) and E.A. Irene (IBM T.J. Watson Research Center). See
     J.D. Meindl et al., "Final Report on Computer-Aided Semiconductor
     Process Modeling", Stanford Electronics Laboratories,
     TR DXG 501-81, July 1981.

(15) H.Z. Massoud, J.D. Plummer and E.A. Irene, IEEE 1981   SISC,
     New-Orleans, Louisiana (1981) (Unpublished).

(16) J.D. Plummer and B.E. Deal, NATO Advanced Study Institute on Process and Device Simulation for MOS-VLSI circuits, Urbino, Italy (1982).

(17) A.I. Ellis, K.M. Gardiner and T.E. Cyr, Electrochemical Society Fall Meeting, Abstr. N° 197, Detroit, Michigan (1982).

(18) P.J. Jorgensen, J. Chem. Phys. 37 (1962) 874.

(19) E. Rosencher, A. Straboni, S. Rigo and G. Amsel, Appl. Phys. Lett., 34 (1979) 254.

(20) A.G. Revesz, J. Electrochem. Soc., 126 (1979) 502.

(21) F.N. Schwettman, K.L. Chiang and W.A. Brown, Electrochemical Society Spring Meeting, Abstr. N° 276, Seattle, Washington (1978).

(22) M. Shatzkes and M. Av-Ron, Thin Solid Films, 91 (1982) 217.

(23) C.M. Osburn and D.W. Ormond, J. Electrochem. Soc., 119 (1972) 597.

(24) E. Harari, J. Appl. Phys., 49 (1978) 2478.

(25) S. Iwamatsu, J. Electrochem. Soc., 129 (1982) 224.

(26) M. Itsumi and F. Kiyosumi, Appl. Phys. Lett., 40 (1982) 496.

(27) P.F. Schmidt, Electrochemical Society Fall Meeting, RNP No 627, Denver, Colorado (1981).

(28) B.H. Yun, Appl. Phys. Lett., 39 (1981) 330.

(29) P.S.D. Lin, T.T. Sheng and R.B. Marcus, Electrochemical Society Fall Meeting, Abstr. No 215, Detroit, Michigan (1982).

(30) C.M. Osburn, J. Electrochem. Soc., 121 (1974) 809.

(31) E.A. Irene, D.W. Dong, R.J. Zeto, J. Electrochem. Soc. 127 (1980) 396.

(32) C. Hashimoto, S. Muramoto, N. Shiono and O. Nakajima, J. Electrochem. Soc., 127 (1980) 129.

(33) E.H. Nicollian, C.N. Berglund, P.F. Schmidt and J.M. Andrews, J. Appl. Phys., 42 (1971) 5654.

(34) S.K. Lai, "Reliability of Thin Gate Insulators", to be published.

(35) B.H. Vromen, Appl. Phys. Lett., 27 (1975) 152.

(36) D.J. Di Maria, D.R. Young and D.W. Ormond, Appl. Phys. Lett., 31 (1977) 680.

(37) R.J. Kriegler, in Semiconductor Silicon 1973, H.R. Huff and R.R. Burgess Eds., The Electrochemical Society Inc., Princeton (1973) 63.

(38) R.R. Razouk and B.E. Deal, J. Electrochem. Soc., 129 (1982) 806.

(39) H.P. Vyas, G.D. Kirchner and S.J. Lee, J. Electrochem. Soc., 129 (1982) 1757.

(40) R.A. Gdula, J. Electrochem. Soc., 123 (1976) 42.

(41) D.R. Young, E.A. Irene, D.J. Di Maria, R.F. De Keersmaecker and H.Z. Massoud, J. Appl. Phys., 50 (1979) 6366.

(42) F.J. Feigl, D.R. Young, D.J. Di Maria, S.K. Lai and J. Calise, J. Appl. Phys., 52 (1981) 5665.

(43) A.G. Revesz, J. Non-Cryst. Solids, 11 (1973) 309.

(44) S.K. Lai, D.R. Young, J.A. Calise and F.H. Feigl, J. Appl. Phys., 52 (1981) 5691.

(45) S.K. Lai and D.R. Young, J. Appl. Phys., 52 (1981) 6231.

(46) G.W. Hughes, R.J. Powell and M.H. Woods, Appl. Phys. Lett., 29 (1976) 377.

(47) G.F. Derbenwick and B.L. Gregory, IEEE Trans. Nucl. Sci., NS-22 (1977) 2151.

(48) J.M. Aitken and D.R. Young, IEEE Trans. Nucl. Scie, NS-24 (1977) 2129.

(49) J.M. Aitken, IEEE Trans. Electron Dev., <u>ED-26</u> (1979) 372.

(50) R.A. Gdula, IEEE Trans. Electron Dev., <u>ED-26</u> (1979) 644.

(51) D.J. Di Maria, L.M. Ephrath and D.R. Young, J. Appl. Phys., <u>50</u> (1979) 4015.

(52) L.M. Ephrath, D.J. Di Maria and F.L. Pesavento, J. Electrochem. Soc., <u>128</u> (1981) 2415.

Infrared Detectors with Integrated Signal Processing

C T Elliott

Royal Signals and Radar Establishment, Malvern, Worcs, England

SUMMARY

The major part of this paper is a review of the operation and characteristics of
a new type of infrared detector for thermal imaging which is known by the acronym
SPRITE (Signal Processing In The Element). In its simplest form the device is a
three-lead structure in n-type cadmium-mercury-telluride which performs the same
function as an array of serial scanned elements together with the associated pre-
amplifiers and time-delay-integration circuits. The operating principle and basic
theory are outlined and compared with experimental data. The application of the
devices to high performance imaging in the 8-14µm band and the potential for lightweight thermoelectrically cooled 3-5µm imagers is described.

A novel approach to electronically scanned "staring" arrays in the 8-14µm waveband
with Cadmium-mercury-telluride detectors will also be presented. The use of n-MOS
silicon switches to address the array has avoided the severe problems associated with
CCD read-out and 8-14µm band imaging has been demonstrated.

1 INTRODUCTION

There are major applications for infrared detectors operating in the 8-14µm and
3-5µm atmospheric transmission bands, principally in thermal imaging and missile
guidance systems. In the majority of cases the longer wavelength band is preferred
because the maximum infrared emission from objects near ambient temperature, and the
maximum differential emission with respect to temperature, occur at around 10µm.
Also, there is better penetration of the atmosphere in this band in conditions of
smoke and mist. The 3-5µm band may be chosen in preference to the long wavelength
band in circumstances where the target temperature is high, or where a smaller
diffraction spread in the optics is needed for better spatial resolution, or when it

is required to reduce the cooling requirements for the detector. The ternary alloy semiconductor cadmium-mercury-telluride (CMT) is now well established as the most important detector material for the 8-14µm band and is increasingly being chosen in preference to other materials, such as InSb, for the short wavelength band.

Infrared systems, unlike visible imaging systems, have employed relatively small numbers of detectors which are scanned over the infrared scene with moving mirrors. The detector arrays used have been restricted to less than about 100 elements because of the problems of making electrical connections to the inner elements of a two-dimensional array and the difficulty of bringing out large numbers of leads from the cryogenically cooled focal plane to ambient temperature. It is the purpose of this paper to describe two new developments by means of which the infrared device performs a signal processing function on the focal plane in addition to a detection function, thereby, reducing the lead-out and interconnect problems.

The main emphasis is on a monolithic CMT device, (1, 2, 3) now known by the commercial name Sprite (4, 5) (Signal PRocessing in The Element), which considerably simplifies current thermal imaging systems and provides a route for future perform-ance improvements through larger areas of CMT on the focal plane. Some recent pro-gress on electronically addressed "staring" arrays operating in the 8-14µm band will also be described. These devices can be used to replace mechanical scanning with electronic scanning in systems where the number of pixels per frame is less than ~ 5000.

2   THERMAL IMAGING WITH DISCRETE DETECTORS

Figure 1 shows schematically some of the detector array configurations and scan formats which have been used in thermal imagers to generate an image which typically contains 10^5 pixels. In a parallel scan system, each detector scans a line in the scene and the outputs are then multiplexed to provide a single video line to a tele-vision display. In a serial scan system, each of the detectors in the row sequen-tially senses the same point in the image. Time-delay and integration (TDI) is per-formed electronically as illustrated to add the signals from the detectors in phase while adding the noise incoherently, thus improving the signal-to-noise ratio by a factor $\sqrt{Q}$ with respect to a single detector, where Q is the number of detectors in the array. A combination of these two scans is known as serial-parallel scanning. In this case the detectors are arranged in a matrix of P parallel by Q serial elements with TDI performed along the rows and multiplexing between the columns. This scan mode is currently favoured by systems designers because it avoids the high mirror rotation rates required for a single row fully, serial system, while providing better channel-to-channel uniformity than a single column, fully parallel system because of the averaging of the detector outputs along the rows.

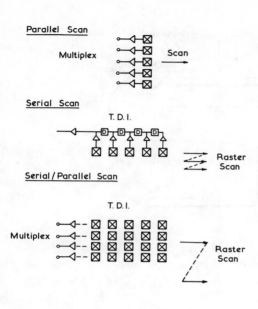

Parallel Scan

Multiplex   Scan

Serial Scan

T. D. I.

Raster Scan

Serial/Parallel Scan

T. D. I.

Multiplex

Raster Scan

Fig 1.   Some of the detector array formats and scanning methods used in thermal imaging.

One measure of the system performance is the noise equivalent temperature difference, ΔT, which may be expressed for any of the scan formats and array configurations (6) as

$$\Delta T = \frac{4F^2 \; B^{\frac{1}{2}}}{A_d^{\frac{1}{2}} tM^* \sqrt{N}} \qquad\qquad 2.1$$

where F is the f/number of the optics, t is the transmission of the optics, B is the electrical bandwidth of the system, A_d is the detector area, $N = P \times Q$ the number of detectors in the array and M^* is a signal-to-noise figure of merit for an individual detector used to compare the performance of detectors with different cut-off wavelengths for the detection of thermal radiation over finite atmospheric paths. It is related to the more familiar figure of merit, D_λ^*, by

$$M^*(T_1,L) = \int_0^\infty D^*(\lambda) A(L,\lambda) \left(\frac{\partial W_\lambda}{\partial T}\right)_{T_1} d\lambda$$

where $A(L,\lambda)$ is the atmospheric transmittance and W_λ is the spectral radiant emittance of a blackbody target of temperature T_1 at range L. It follows that the parameter $M^* \sqrt{N}$ can be regarded as a figure of merit for a detector array used in thermal imaging and we will use the concept later of an effective number of elements, N_{eff}, in comparing the performance of Sprite detectors with discrete detector arrays.

3   OPERATING PRINCIPLE OF THE SPRITE DETECTOR

The device structure in its simplest form for a single-row serial system is illustrated in Figure 2a. It consists of a strip of n-type CMT whose dimensions are determined by the system optics and scan speed, but they may be for example, 1mm long, 50μm wide and 10μm thick. A constant current circuit provides bias through the two ohmic end contacts and the third contact, also ohmic is a potential probe for signal read-out. Alternatively, a reverse biassed p-n junction could be used for read-out, but the technology for p on n diodes in CMT is not well established. The bias field is sufficiently high for excess minority carriers generated optically over a significant fraction of the filament length to reach the negative end contact before recombination occurs in the bulk i.e. the ambipolar transit time along the strip is comparable to or less than the excess carrier lifetime τ. Further the bias

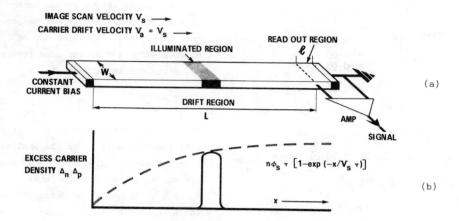

Fig 2.   *The operating principle of a Sprite detector. The upper part of the figure shows a CMT filament with three ohmic contacts. The lower part shows the build-up of excess carrier density in the device as a point in the image is scanned along it.*

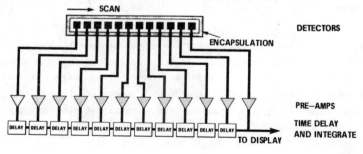

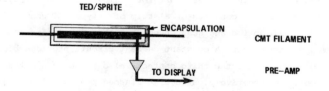

Fig 3.   *How Sprite simplifies the electronics of a thermal imager. The upper part of the figure shows one row of conventional detectors from a serial-parallel scan imager and the associated electronics to perform time-delay integration. The lower part shows a Sprite detector performing the same function*

field is selected so that the ambipolar drift velocity v_a is equal to v_s, the image
scan velocity. v_a is the product of the ambipolar mobility μ_a and the bias field ε,
where

$$\mu_a = \frac{(n-p)\ \mu_n \mu_p}{n\mu_n + p\mu_p} \qquad\qquad 3.1$$

with n and p the electron and hole densities respectively, and μ_n and μ_p the electron
and hole drift mobilities. In n-type CMT where $\mu_n \gg \mu_p$

$$\mu_a \simeq \left(1 - \frac{p}{n}\right)\mu_h \qquad\qquad 3.2$$

In most practical situations $n \gg p$ and μ_a approximates to μ_p.

Consider now an element of the image scanned along the strip as illustrated by the
shaded region of Fig 2a. The excess carrier density in the strip, at a point corresponding to the illuminated element, increases during the scan as illustrated in
Fig 2b. When the illuminated region enters the read-out region, between the potential probe and the negative contact, the increased conductivity modulates the voltage on the read-out contact and provides an output signal. Since the integration
time of the signal flux, which is $\sim \tau$ in a long element (or L/v_s in a short element)
can be greater than the dwell time for a point in the image on a conventional discrete element in a fast-scanned, serial-parallel system a better signal-to-noise
ratio is obtained. In a background limited detector the principal noise is fluctuations in the excess carrier density generated by the background flux. The background
is also integrated for a period $\sim \tau$, but since the uncertainty in the excess carrier
density is proportional to the square-root of the integrated flux a net gain in signal-to-noise ratio with respect to a discrete element is obtained which is proportional to $\tau^{\frac{1}{2}}$. The simple three-lead structure with a single preamplifier, therefore,
performs a detection and time-delay-integration function which previously required a
row of discrete detectors with their associated amplifiers and time delay circuits.
The simplification in the system which results from the replacement of a serial row
of conventional detectors by a Sprite is illustrated in Figure 3.

Another advantage of the Sprite is the very high levels of signal output voltage
or responsivity compared to conventional photoconductive detectors. This is due to
the relatively larger conductivity modulation resulting from the integration period
of $\sim \tau$, whereas in a small photoconductive detector the integration period is
approximately the minority carrier transit time across the device.

The material requirements for the device are determined by the need for long
excess carrier lifetime to provide long integration and low minority carrier diffusion
length, ℓ_h, in order to provide good spatial resolution. The subject of spatial
resolution is discussed in detail below but at this point we may note that the

TABLE 1   PROPERTIES OF n-TYPE CMT

Waveband	8-14μm	3-5μm	
Operating Temperature (K)	80	190	230
τ (μs)	2(5)	15(30)	15
μ_h ($cm^2 v^{-1} s^{-1}$)	480	~150	~100
D_h ($cm^2 s^{-1}$)	3.2	2.5	2.0
ϱ_h (μm)	25	61	55

resolution size as determined by carrier diffusion is normally limited to approximately $2\varrho_h$. The list of material properties for n-type CMT shown in Table 1 show this to be a very suitable material for the device and perhaps uniquely suitable for 8-14μm band operation. Lifetime values of ~ 2μs are currently obtained in alloy compositions suitable for operation in this band at 77K. Recombination is believed to be dominated by an intrinsic Auger process in these compositions and lifetimes up to ~ 5μs (achievable in low background conditions) may be obtained from lower carrier density material. Lifetime values of 10-20μs are observed in compositions used for 3-5μm band operation at temperatures achievable with thermoelectric coolers and values ~ 30μs have been observed (7). The low hole mobility in CMT results in short diffusion lengths even with the long lifetimes, yielding a spatial resolution in the 8-14μm band Sprite devices similar to that normally expected from discrete devices. The spatial resolution in the 3-5μm band of $2\varrho_h$ is poorer than that normally required from discrete devices but this may be corrected either by limiting the integration period to less than τ or by means of a meander path structure as discussed below.

Another material which would be suitable for 3-5μm band, 77K operation is InSb in which $\mu_a \simeq 425$ and lifetimes of 7-10μs have recently been reported by Pines and Stafsudd (8).

4   SPRITE FABRICATION AND DEVICE GEOMETRY

Sprite devices are fabricated from Bridgman grown n-type CMT with extrinsic carrier concentrations in the range 3 to 5 x 10^{14} cm^{-3} and with other properties as shown in Table 1. The earliest devices made were single or double filaments in which the read-out contact was fabricated, using photolithographic techniques, by means of a gold overlay going from the substrate in a small triangle over the edge of the filament. Eight, sixteen or twenty-four row devices in a close-packed structure are now fabricated by Mullard using an elegant monolithic process. The CMT slice is polished to its final thickness of 8μm and divided into chips, typically 1.5mm x 0.5mm

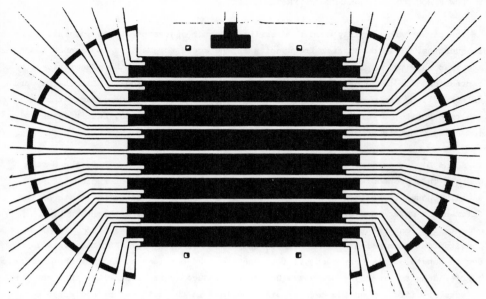

Fig 4. Photograph of an eight-row Sprite. Courtesy of Mullard Ltd.

in size for fabrication of an eight-row Sprite. The CMT chips are glued onto
sapphire substrates, contact metallization is applied and the array is defined photo-
lithographically. The individual elements are then formed by etching. A photograph
of an eight-row Sprite produced in this way is shown in Figure 4. The edge of the
original chip of CMT is shown as the oval outline. The Sprite filaments which appear
black are 700µm long and 62.5µm wide. The ends of the filaments are forked (known as
bifurcated read-outs), the wider prong of the fork carrying the bias current, the
narrower prong forming the read-out probe. The read-out dimensions are 50µm long by
38µm wide. The mid-grey areas of the photograph show the gold metallization running
out and over the edge of the monolith onto the sapphire carrier, from which it is
bonded to leads in the encapsulation. Although read-outs are shown at both ends of
the Sprite filaments the set at one end only is used.

The bifurcated type of read-out is adopted, in preference for example to a side-
arm at right angles to the filament, because it permits a close packed two-dimensional structure. A two-dimensional array of conventional detectors must be
arranged in a staggered format to permit connections to be made. Such an array can
be less efficiently cold-shielded and needs the additional complexity of delays in
the system electronics to compensate for the element displacements.

5 LOW FREQUENCY PERFORMANCE OF SPRITE DETECTORS

The parameters normally used to describe detector operation at low modulation frequencies, i.e. low spatial frequencies in the scene for devices in scanned imagers, are responsivity, R, detectivity, D^*, and noise per unit bandwidth. Theoretical expressions for these parameters in Sprite detectors are derived by Elliott, Day and Wilson (2) and are re-stated below. D^* in particular is not an ideal parameter to describe Sprite performance because it is dependent on scan speed, but it is useful in making comparisons with discrete devices.

R and D^* are defined with respect to a device size W x W, where W is the device width. This procedure is adopted since the length of the read-out region, ℓ is usually chosen to be less than W and the spatial resolution along the length of the filament is determined by the diffusive spread of the carriers rather than by geometrical constraints. Normally the width of the device is chosen to match the longitudinal resolution. Other assumptions which are made are: that the overall device length, L, is very much greater than the read-out length, ℓ; that the extrinsic carrier density greatly exceeds the background induced carrier density; that the effects of diffusion on the carrier distributions in the bulk may be neglected at low modulation frequencies; that the positive bias contact is non-injecting and that the effective recombination velocity of the negative contact is equal to the minority carrier drift velocity in the filament. The last assumption cannot be satisfied in practice but the error introduced is small if the recombination velocity is higher than the drift velocity, the excess carrier density being reduced below the calculated values over an upstream diffusion length. A more serious error is introduced if the contact is blocking to minority carriers when a large excess density may be produced by accumulation (9, 10, 11).

The responsivity for monochromatic radiation of wavelength λ is given by

$$R_\lambda = \frac{\eta\tau\varepsilon\ell}{E_\lambda W^2 tn} \left[1 - \exp(-L/\mu_a\varepsilon\tau) \right] \tag{5.1}$$

where η is the quantum efficiency, ε is the bias field, E is the photon energy, t is the device thickness and n is the equilibrium electron density. In a device which is long compared to an ambipolar drift length, $L \gg \mu_a\varepsilon\tau$

$$R_\lambda = \frac{\eta\tau\varepsilon\ell}{E_\lambda W^2 tn} \tag{5.2}$$

The dominant noise source is generation-recombination noise due to fluctuations in the density of thermally generated and background radiation generated carriers and the spectral density is given by

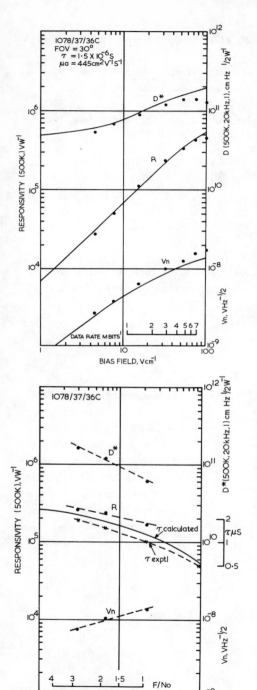

Fig 5.   Performance parameters of an 8-13μm band Sprite operated at 77K in 30° field of view.

Fig 6.   Performance of an 8-13μm band Sprite operated at 77K as a function of background flux at a data rate of 3 x 10⁶ pixels/second.

$$v_n^2 = 4 \frac{\varepsilon^2}{n^2} \frac{\ell\tau}{Wt} \left(p_o + \frac{\eta\phi_b\tau}{t} \right)[1 - \exp(-L/\mu_a\varepsilon\tau)] \left[1 - \frac{\tau}{\tau_a} \ 1 - \exp(-\tau_a/\tau) \} \right] \qquad 5.3$$

where p_o is the free hole density in thermal equilibrium, τ_a is the transit time through the read-out zone ($\ell/\mu_a\varepsilon$) and ϕ_b is the background flux density.

The detectivity may be obtained from equations 5.1 and 5.3. For a long, background device in which $L \gg \mu_a\varepsilon\tau$ and $\eta\phi_b\tau/t \gg p_o$

$$D_\lambda^* = \frac{\eta^{\frac{1}{2}}}{2E_\lambda} \left(\frac{\ell}{\phi_b W} \right)^{\frac{1}{2}} \left[1 - \frac{\tau}{\tau_a} \left\{ 1 - \exp(-\tau_a/\tau) \right\} \right]^{-\frac{1}{2}} \qquad 5.4$$

and at sufficiently high scan speeds such that $\tau_a \ll \tau$

$$D_\lambda^* = (2\eta)^{\frac{1}{2}} D_\lambda^*(\text{BLIP}) \left(\frac{\ell}{W} \right)^{\frac{1}{2}} \left(\frac{\tau}{\tau_a} \right)^{\frac{1}{2}} \qquad 5.5$$

It is often convenient to express D_λ^* in terms of the pixel rate, S, which for a nominal resolution size of W x W is ν_s/W. Then in the high scan speed limit

$$D_\lambda^* = (2\eta)^{\frac{1}{2}} D_\lambda^*(\text{BLIP}) \times (S\tau)^{\frac{1}{2}} \qquad 5.6$$

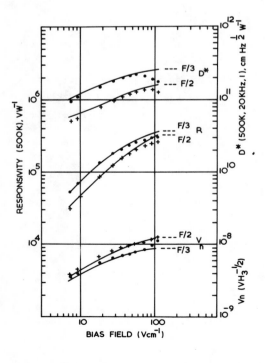

Fig 7. *Performance of a bifurcated Sprite with a cut-off wavelength of 11.5μm operating at 77K. Results are shown for cold shield apertures corresponding to F/2 and F/3. The solid lines are calculated using τ = 2.0μs, μ_a = 432cm² $v^{-1}s^{-1}$ for F/3 and τ = 1.3μs, μ_a = 412cm² $v^{-1}s^{-1}$ for F/2. η = 0.6.*

The device yields useful improvements in detectivity relative to a background limited discrete device when the value of $S\tau$ exceeds unity. For CMT material with alloy composition appropriate to 8-14µm operation at 77K (x = 0.205), typical values of τ are typically ~ 2µs and useful performance is obtained in serial or serial-parallel scan imaging systems with channel pixel rates exceeding ~ 5×10^5 s^{-1}. For alloy compositions used in the 3-5µm band at thermoelectric temperatures, τ values greater than 15µs can be obtained and useful performance can be obtained at channel pixel rates exceeding ~ 10^5 s^{-1}. A further interesting property of the device, which follows from equation 5.5 is that the signal-to-noise ratio is measured in the total system bandwidth is independent of scan speed and data rate, provided that the device is sufficiently long for the condition $L \gg \nu_s \tau$ to be maintained.

The performance can alternatively be expressed in terms of the number of background limited photoconductive elements needed in a serial array, N_{eff}(BLIP), to give the same signal-to-noise ratio - see Section 2. Since the effective D_λ^* of a serial array increases as the square-root of the number of elements, it follows from equations 5.4 and 5.6 that

$$N_{eff}(\text{BLIP}) \;=\; \frac{\ell}{W} \left[1 - \frac{\tau}{\tau_a} \left\{ 1 - \exp(-\tau_a/\tau) \right\} \right]^{-1} \qquad\qquad 5.7$$

and when $\tau \gg \tau_a$

$$N_{eff}(\text{BLIP}) \;=\; 2S\tau \qquad\qquad 5.8$$

The discrete elements have been assumed to have the same quantum efficiency as the Sprite. The factor 2 is replaced by 3/2 when comparing with ideal photoconductor elements in sweep-out, and unity when comparing with ideal photovoltaic elements in reverse bias. In practice Sprite devices are found to behave in a very ideal way as the background radiation is reduced and they can be fabricated in very compact, well cold shielded arrays. Consequently they are equivalent to larger numbers of typical discrete elements than equations 5.7 and 5.8 would suggest, particularly in narrow fields of view.

Measurements of the detector parameters at modulation frequencies below the high-frequency roll-off of the device can be made with the same techniques normally used for discrete devices. The responsivity has been measured by exposing the whole device to chopped but unscanned radiation from a 500K blackbody source.

An example of the results obtained from one of the early devices fabricated by Mullard, with the gold overlay type of read-out, are shown in Figure 5 and Figure 6. The filament is 1mm long, 50µm wide with a 65µm long read-out and operates in the 8-14µm band at 77K. In Figure 5, D^*, responsivity and noise are plotted as a function of bias field for a case where the background flux was defined by a 30° field-

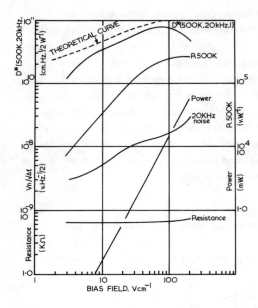

*Fig 8.　Performance parameter for a
Sprite with cut-off wavelength
of 4.5μm operated at 183K in
55° field of view.*

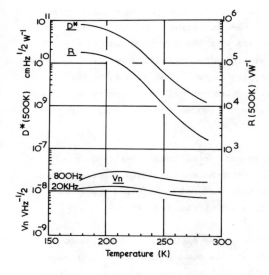

*Fig 9.　3–5μm band Sprite performance
versus temperature. Measured
with an F/1 equivalent cold
shield aperture and a bias
field of 40V cm⁻¹.*

of-view circular cold shield. The solid lines shown on the graph are computed from equations 5.1, 5.3 and 5.4 using the measured value of the excess lifetime 1.5μs and a value of ambipolar mobility of 445cm^2 v^{-1} s^{-1} calculated from equation 3.1. The D_λ^* values are reduced from peak values to 500K blackbody values for comparison with experiment. The device is unbloomed and the quantum efficiency is taken as 0.6. It carries a native oxide passivation layer to minimise surface recombination which accumulates the surface and results in a measured resistance which is approximately half that which would be predicted from the measured extrinsic concentration of 5×10^{14} cm^{-3}. This effect has been allowed for by adjusting the value of n used in the equations until a good fit to the responsivity plot is obtained. The value used was 1×10^{15} cm^{-3} and also gave good fits to data obtained with different levels of background flux (2).

The agreement between the experimental and theoretical plots is good except at high bias levels, greater than about 50V cm^{-1}, where the temperature of the device is raised by Joule heating. It may be noted that in addition to high D* values, very high values of responsivity are obtained compared to a discrete photoconductor, for which sweep-out effects limit the maximum value to about 2×10^4V W^{-1}.

Plots of the detector parameters for the same device as a function of background flux at a data rate of 3×10^6 pixels s^{-1} are shown in Figure 6. Background limited operation is demonstrated down to F/3 field of view. Good cold-shielding is impor-tant in achieving good detectivity since the background flux determines not only the noise in the device but also affects the lifetime and hence the integration time. The lifetime im material with alloy composition appropriate to 8-14μm band operation at 80K is limited by an Auger process.

An example of the results obtained by Wotherspoon (12) from a bifurcated Sprite of the type shown in Figure 4 is given in Figure 7. Plots of the detector parameters as a function of bias field are shown for cold-shield apertures equivalent to F/2 and F/3 field of view. The solid lines are computed from equations similar to 5.1, 5.3 and 5.4 modified to include the effect of the bifurcated read-out. The material parameters used in the calculation are shown in the figure caption.

An example of Sprite operation in the 3-5μm band at thermoelectric temperatures is given in Figure 8. The device has a cut-off wavelength of 4.5μm and is operated on a four stage thermoelectric cooler at 183K. The filaments are 1mm long and 50μm wide with a read-out length of 35μm and an excess carrier lifetime of approximately 10μs. The detectivity and responsivity are 500K blackbody values. The tendency to saturation in the responsivity and the noise can be seen when the integration becomes limited by the finite length of the filament. The theoretical plot shown for com-parison was calculated using equation 5.4 with $\eta = 0.6$ and $\mu_a = 130$cm^2 v^{-1} s^{-1}.

3-5μm band devices have been shown to operate satisfactorily at temperatures which are achievable with two-stage Peltier coolers. This has considerable significance for simple lightweight imaging systems because of the improved efficiency and reduced input power requirements of Peltier coolers as the number of stages is reduced. Results obtained by Blackburn et al (4) on 3-5μm band Sprite parameters as a function of operating temperature are reproduced in Figure 9 and show that D^* (500K) greater than 1×10^{10} cm Hz$^{\frac{1}{2}}$ W^{-1} may be obtained at temperatures up to 240K. The results are from one element of an eight row device which is 700μm long.

6   HIGH FREQUENCY PERFORMANCE AND SPATIAL RESOLUTION OF SPRITE DETECTORS

The response of the device to high spatial frequencies in the image is limited fundamentally either by spatial averaging due to the finite size of the read-out zone or by the diffusive spread of photogenerated carriers in the filament. The response may be further degraded by imperfect matching of the carrier drift velocity to the image velocity. Day and Shpeherd (13, 14) show that the detector transfer function due to diffusion and velocity mismatch, when $L \gg \nu_s \tau$, is given by

$$\text{DTF}\big|_{\text{diffusion}} = \frac{\exp(-i\phi)}{\left[(Q_a^2 k_s^2 + 1)^2 + \left\{ (\mu_a \epsilon - \nu_s) k_s \right\}^2 \right]^{\frac{1}{2}}} \tag{6.1}$$

where

$$\phi = \tan^{-1}\left[k_s \tau (\mu_a \epsilon - \nu_s) / (Q_a^2 k_s^2 + 1) \right] \tag{6.2}$$

and k_s is spatial frequency in the image.

Plots of equation 6.1 for cases of practical interest show the modulus to be insensitive to velocity mismatch of less than about 5% and this is confirmed experimentally. In the matched condition when $\mu_a \epsilon = \nu_s$, $\phi = 0$ and the overall modulation transfer function, including the spatial averaging of the read-out, is given by

$$\text{MTF} = \left(\frac{1}{1 + k_s^2 Q_a^2} \right) \left(\frac{2\sin k_s \ell/2}{k_s \ell} \right) \tag{6.3}$$

Since the signal-to-noise ratio for a Sprite operated at high scan speed is independent of the read-out length it is advantageous in this situation to choose ℓ sufficiently small for the diffusion term to dominate the MTF. An alternative which maintains the responsivity is to reduce the width of the device in the read-out zone thereby increasing the bias field locally and reducing τ_a as in the bifurcated read-out geometry of Figure 4.

An example of the MTF measured on a filament from an eight row bifurcated Sprite by Braim and Davis (15) is shown in Figure 10. The experimental result is compared

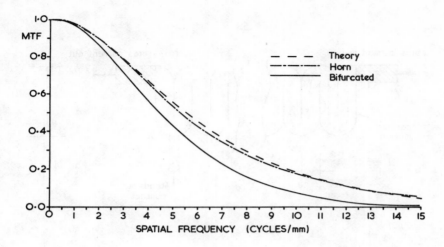

Fig 10.   *The modulation transfer function measured on bifurcated and horn read-out*
(9), 8–14μm Sprites.   The theoretical curve is calculated from equation
6.3 with Q_a = 25μm and l = 50μm.   (After Braim and Davis (15)).

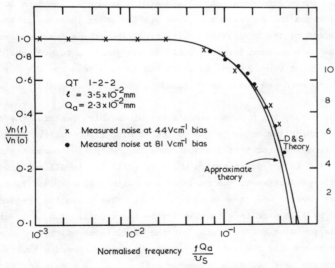

Fig 11.   *Noise voltage spectrum of an 8–13μm band Sprite at 77K.*

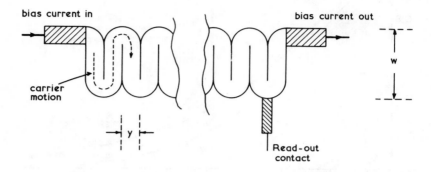

Fig 12.   Schematic illustration of a meander-path Sprite.

with a calculated curve from equation 6.3 assuming a read-out length of 50µm. The
lower values of MTF on the experimental curve are due to an increase in the effective read-out length of the bifurcated device resulting from a carrier transit-time
spread due to the non-uniform fringing field at the junction between the drift
region and the read-out zone. Also, shown on Fig 10 is a measured MTF for a device
with a modified, horn-shaped read-out designed to reduce the transit time spread (9).

Day and Shepherd (13, 14) have obtained expressions which describe the behaviour
of generation-recombination noise in the device at high frequencies. Implicit in
their model is the assumption that a disturbance in the carrier concentrations,
occurring at a point in the drift region which is initially localized in space and
time, subsequently propagates to the read-out zone with a line-spread function
obeying the normal macroscopic transport equations. Thus the noise frequency spectrum is modified by diffusion in a similar way to the signal. The topic of diffusion-
modified noise is somewhat controversial but the predictions of the model are consistent with experimentally observed behaviour. Day and Shepherd's equations are
quite complex and are best suited to computer evaluation. A simple empirical expression which is found to give a reasonably good fit to experimental data on long
devices is

$$V_n(f) \; = \; V_n(m) \left[1 \, + \, (2\pi f Q_a / V_s)^2 \right]^{-\frac{1}{2}} \; \mathrm{sinc}(\pi f \ell / v_s) \qquad\qquad 6.4$$

where $V_n(m)$ is the rms value of the low frequency generation-recombination noise
voltage given by equation 5.3. An example of the noise spectra obtained from an

8-14 μm band device operating at 77K is shown in Figure 11. Theoretical plots from equation 6.4 and from Day and Shepherd's model are shown for comparison.

Since both MTF and noise roll-off at high frequencies there is some advantage in practice in employing an amplifier with high frequency lift. The form of the graphs at high frequencies is different from that for conventional discrete devices, but comparison of the signal/noise ratio for a short read-out Sprite with a discrete device indicates that the Sprite has an equivalent information bandwidth to a discrete device ∿2.2 Q_a long. Thus for 8-14 μm band 77K operation the resolution size of a long device is ∿55 μm. There is some advantage, therefore, in increasing the width of the devices from say 50 to 75 μm and scaling the focal length of the detector lens in the imaging system in the same ratio, keeping the diameter constant. Provided that the cold shield aperture of the detector is matched to the new f/No of the lens, there is no loss of signal-to-noise performance since the devices remain BLIP limited in low field of view. This procedure has the additional benefit of reducing the background flux density on the detector and of increasing the excess carrier lifetime. A disadvantage of scaling the size by a factor S is to increase the power dissipation in the detector by a factor S^4. Though not serious for the relatively small arrays currently used, it is desirable to minimise dissipation in the larger arrays projected for future systems. Since it is necessary to improve resolution in the direction perpendicular to the filaments a saving in power would be obtained by scaling the image, with anamorphic optics, in the filament direction only. Joule heating is then increased as S^3. This technique is discussed in more detail by Sleigh (16).

For intermediate temperature operation of 3-5 μm devices the spatial resolution of a long device is ∿140 μm which is normally unacceptable and the technique of size scaling is less satisfactory than for the 8-14 μm band because the devices operate closer to thermal g-r noise limited performances. One method to improve the resolution is to use a short device where the transit time in the filament is less than the lifetime in order to limit the diffusive spread. An alternative which improves the spatial resolution without loss of integration, and hence signal-to-noise performance is the meander-path structure (17) shown in Figure 12. The bias field is chosen such that mean velocity of carriers in a direction parallel to the image scan is matched to the image velocity. If we consider a line in the image scanning along the device the effective diffusion length of carriers about this line is reduced by a factor W/Y. It may be shown that the MTF for such a structure is given approximately by

$$\text{MTF} = \text{sinc}^3 \frac{kY}{2} \left[1 + \left(\frac{kQ_a Y}{W} \right)^2 \right]^{-2} \qquad\qquad 6.5$$

An experimental demonstration of the improved spatial resolution from a device of this type is reported by Blackburn et al (4). The power dissipation penalty of this structure is less than to that for scaling with anamorphic optics being proportional to $(W/Y)^2$.

7   SYSTEMS USE OF SPRITE DETECTORS AND FUTURE STRETCH POTENTIAL

Parallel arrays of Sprites can be substituted for discrete element arrays in serial/parallel imaging systems when the pixel rate per channel exceeds about $5 \times 10^5 \text{ s}^{-1}$ for 8-14 μm band operation and 10^5 s^{-1} for 3-5 μm operation. For a given system bandwidth the signal-to-noise performance is relatively insensitive to the number of parallel elements provided that the channel bandwidth is sufficiently high for significant integration in the Sprites. The choice of devices is therefore largely a systems trade-off between high rotor speeds and very simple electronics.

Systems operating in the 8-14 μm band have to date used 2 row, 4 row and 8 row devices. The performance of the Sprite arrays in these systems is equivalent to that which would be obtained from an array of about 50-100 BLIP discrete detectors. In practice the good cold-shielding which is possible on the compact Sprite arrays and their near-ideal performance means that somewhat larger numbers of conventional devices are required to give equivalent performance. The principal benefit therefore of Sprite to date in systems has been in allowing considerable simplification of the electronics. Although Sprite detectors can be substituted directly into serial/ parallel systems designed for conventional detectors there are performance advantages to be gained from designing the system to exploit the characteristics of the Sprite. Some of these system design considerations have been discussed in papers by Sleigh (16), Webb and Braim (18), Runciman and MacGregor (19) and Cuthbertson and MacGregor (20). An example of a thermal image obtained with an 8 row Sprite operating in the 8-14 μm band is shown in Figure 13. It is photographed from the TV monitor of an indirect-view thermal imaging system manufactured by Marconi Avionics Ltd and Rank Taylor Hobson with the support of the UK Ministry of Defence for the Thermal Imaging Common Modules programme.

Sprite detectors have mainly been used in high performance 8-14 μm systems so far, but they have potentially important future applications in simpler, lighter systems employing thermoelectric cooling and operating in the 3-5 μm waveband. Moore and Barringer (21) have described an experimental imager of this type weighing under 5.4 kg and consuming 12 Watts of power from throw-away batteries. They predict that these figures will be halved in development. A design for a lightweight imager based on Sprite has also been described by Bleicher et al (22). An example of a thermal image is shown in Figure 14 obtained from an experimental imager built by PRL and MEL and using an 8 row Sprite operating at 190K on a thermoelectric cooler. It is a direct view system and the image shown was obtained by photographing the

Fig 13.   *Photograph of an 8-14 μm thermal image in black hot mode, obtained*
           with an eight filament device (Courtesy Marconi Avionics Ltd and
           Rank Taylor Hobson).

Fig 14.   *Photograph of a 3-5 μm thermal image in white hot mode obtained from an*
           eight filament Sprite operating on a thermoelectric cooler.
           (Courtesy of M Martin, RSRE).

light-emitting diode display.

   The devices also have considerable potential for stretch to very much larger
arrays in both wavebands with relatively straightforward extensions of the well
established photoconductive detector technology. Parallel arrays of 16 and 24
elements have already been fabricated and two-dimensional arrays will be developed
in future with time-delay-integration cricuitry re-introduced between Sprite devices
along the serial rows. Two particular features of the Sprite which facilitate the
fabrication of close-packed, two-dimensional structures, which may be hybridized
with silicon circuitry in the larger sizes, are; the order of magnitude reduction
in the number of interconnects required compared to conventional devices and the
high responsivity levels which are compatible with low power consumption buffer
amplifiers. The ultimate size of such arrays is limited by the heat-load imposed
by Joule heating but it is expected that equivalent performance to an array of more
than 1000 BLIP elements in the 8-14 μm band will be achieved with a heat-load of
less that 1 Watt. Arrays equivalent to hundreds of conventional elements should be
possible in the 3-5 μm band within the constraints of available cooling power from
Peltier coolers.

8   ELECTRONICALLY SCANNED DETECTOR ARRAYS FOR 8-14 μm BAND OPERATION

   There is a large research activity currently directed towards two-dimensional
arrays of detectors consisting of 1000 to 5000 elements which are addressed
electronically. These "staring" arrays will find applications in a new generation
of sophisticated missile homing heads and perhaps also in simple, small lightweight
imagers. Two major benefits will result from the use of the new devices. The first
is the elimination of mechanical scanning, which will reduce production costs, reduce
the size of systems and improve reliability. The second is the increased
sensitivity due to the longer integration period of the detectors when compared with
a scanned system operating at the same frame-rate. Alternatively, the increase in
sensitivity may be traded for reduced optical apertures and decreased system size.

Many different technologies for electronically scanned arrays are being investigated,
almost all of which utilise charge transfer devices (CTDs) to multiplex the
information from the detectors. In the hybrid scheme, an array of photodiodes
fabricated in a narrow-gap semiconductor, commonly CMT, is interconnected with a
siliton substrate containing CTDs. There are two major problems in this approach
for 8-14 μm band operation; the first associated with the difficulty of injecting
charge into the CTD, and the second with the limited charge storage capacity of
these devices. Figure 15 shows schematically the coupling of an infrared diode to a
CCD for direct injection and the equivalent circuit. The input circuit behaves like
a grounded gate MOS transistor, the input diffusion acting as a source and the first

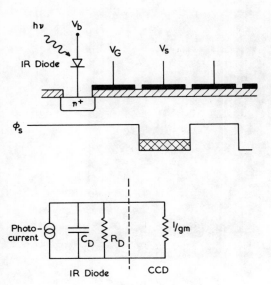

Fig 15.
Schematic diagram and equivalent
circuit of an infrared diode
injecting charge directly into a
CCD through an input diffusion.

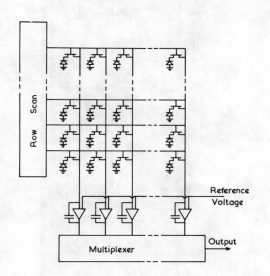

Fig 16.
Schematic diagram of a line addressed
array.

Fig 17.   *32 x 32 array of cadmium-mercury - telluride photodiode on a 40 μm
pitch interconnected with MOSFET switches in a silicon substrate.
(Courtesy of Mullard Ltd).*

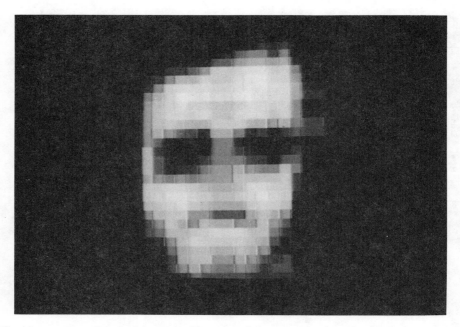

Fig 18.   *Example of electronically scanned imagery showing the face of a man
with glasses and beard.*

CCD well as a virtual drain. For efficient injection of charge the photodiode is required to have an impedance greater than that of the CCD input. The trans-conductance, g_m, of the CCD input is determined by the background generated current flowing from the IR diode and for the typical background current in the 8-14 μm band, will be close to the weak inversion region such that $g_m \simeq \frac{qI}{\alpha kT}$, where α lies between about 3 and 5. It may be shown for example that to achieve an 80% injection efficiency from a diode with cut-off wavelength of 10.5 μm in a background flux of 2.5×10^{14} cm^2 s^{-1}, a zero-bias-resistance times area product ($R_o A$) for the diode of more than about 5 is required, which is higher than the currently achieved in routine fabrication of two-dimensional diode arrays. The charge capacity of the CTD is a problem because of the high level of background flux in the 8-14 μm band from near ambient temperature objects. The capacity of a silicon CTD is typically about 5×10^{11} cm^{-2}. This means that saturation of the storage wells occurs in a period of only 20 μs in a photon flux of 2.5×10^{16} cm^2 s^{-1}, which is approximately the photon flux density absorbed by a detector with a cut-off wavelength of 10.5 μm and quantum efficiency of 0.8 in F/2 field of view. If adequate integration time is to be achieved at practicable frame rates from a hybrid CTD, background subtraction circuitry must be developed to remove part of this pedestal charge, which is sufficiently small to incorporate in the substrate within a small fraction of the pixel area. In an alternative monolithic approach using narrow-gap semi-conductors, the injection problem is avoided, since the carriers are generated optically in CTD wells formed in the narrow-gap material, but the well capacity problem is exacerbated. CMT CTDs have been fabricated for operation in the long-wavelength band (see Beck et al (23)). The well capacity in these devices is limited by band-to-band tunnelling to less than 10^{11} cm^{-2} and the long-wavelength cut-off of the devices has so far been limited to about 9.2 μm. Monolithic devices are also fabricated in silicon using extrinsic photoconductors and CTDs, but the low operating temperatures required make the device unsuitable for many applications.

In a collaborative research activity RSRE and Mullard have demonstrated a relatively simple method of achieving electronically scanned arrays in which the multiplex operation is carried out by n-MOS switches. This has been described previously in papers by Ballingall (24), Baker et al (25) and Ballingall et al (26). The switch approach, as will be shown below, will realise much of the expected advantages of electronically scanned arrays while avoiding the severe problems associated with charge transfer devices.

A schematic diagram of the arrangement of a line-addressed N x N array is shown in Figure 16. Each infrared diode has its anode connected to a common ground line and the cathode connected through a switch to an output rail. The control gates of the switches in each row are connected together and addressed by a row select shift register. In operation, each row of diodes in turn is selected by switches and connected to integrating amplifiers, the outputs from the latter being multiplexed to provide a single video line. Since there is no storage at the detectors, integration occurs only for a line sample time t_f/N, where t_f is the frame time. In this arrangement the signal/noise performance of the array is equivalent to that which would be obtained from a linear array of N detectors mirror scanned across the scene. The R_oA required for the diodes using this scheme is not difficult to achieve. For example, using an amplifier with a JFET input with an equivalent input noise voltage of 5 nV Hz$^{-(\frac{1}{2})}$, a diode resistance of $> 4K \Omega$ is required for the amplifier to degrade the detectivity of the diode by less than 3 dB. This corresponds to an $R_oA \simeq 0.1 \Omega$ cm^2 for a diode area of 50 µm square. The circuit is capable of very large dynamic range and the background pedestal is removed very simply by subtracting a reference voltage at the amplifier input.

Figure 17 shows a 32 x 32 array of CMT photodiodes on a 40 µm inter-element pitch connected with MOSFET switches in a silicon substrate. Two shift registers in the substrate allow either single-element addressing for test purposes or line sampling as shown in Figure 16. The on resistance of the switches is less than 1K Ω at the operating temperature of 77K, and the associated Johnson noise is small compared with the detector noise. The integrating amplifiers have not yet been incorporated on the focal plane and 32 output leads are taken from the cryogenic encapsulation. The results given below are from an array with a cut-off wavelength of 12 µm and with all 1024 elements connected. The mean R_oA of the detector is 1.0 Ω cm^2 and 99.5% had $R_oA > 0.1$ at the operating temperature of 77K.

A potential problem when multiplexing low level signals with switches is the occurrence of spurious signals due to switching transients. These transients have been found to have a duration of less than 20 µs and they can be avoided by sampling the signal. The loss of integration time and signal-to-noise performance is small, even at frame rates of several hundred Hertz.

Because the switches and transimpedance amplifiers have resistances which are very small compared with the detector slope resistance, the nonuniformity of the array is entirely due to variations in the current responsivity of the diodes. For the array shown, the standard deviation over the mean responsivity was 0.36. Uniformity correction was performed by microprocessor-based off-focal-plane electronics, which comprised data converters combined with digital memory and logic. Correction factors for variations in responsivity and background signal were stored in memory. These were derived by exposing the array to a uniform scene at two different temperatures. The uniformity was improved in this way by about a factor of 20. An example of the imagery which has been obtained at this early stage is shown in Figure 18. This shows the face of a man with glasses and a beard, obtained using an F/1 lens and with no cold shield on the detectors. The estimated noise-equivalent temperature difference is about O.7K and is limited by residual fixed pattern noise. The detectivity of elements in a similar array was measured using a 5OOK blackbody source and gave values for D* (5OOK), without cold shielding, in the range $1 - 1.5 \times 10^{10}$ cm $Hz^{\frac{1}{2}}$ W^{-1}.

The line addressed array shows considerable promise for realising one of the advantages of staring arrays, that of no mechanical scanning. As described the sensitivity will be no greater than that of a mirror-scanned linear array, but a closer approach to the sensitivity advantages of a true staring array may be obtained by reading out many lines from the device simultaneously. The scheme also offers the possibility of random access to required parts of the scene. By replacing the shift registers shown in Figure 17 by decoders, for example, any group of 32 diodes can be selected.

ACKNOWLEDGEMENTS

The author gratefully acknowledges the contributions to this work of colleagues both at RSRE and Mullard, too numerous to mention individually. Part of the work described was supported by the UK Ministry of Defence, Procurement Executive, DCVD.

REFERENCES

(1)   C.T. Elliott, Electronics Letters 17 No 8 312 (1981)

(2)   C.T. Elliott, D. Day and D.J. Wilson, IR Physics 22 31 (1982)

(3)   C.T. Elliott, Proc. Int. Conf. on Advanced IR Detectors and Systems, IEE Conference publication No. 2O4 1 (1981)

(4)   A. Blackburn, M.V. Blackman, D.E. Charlton, W.A.E. Dunn, M.D. Jenner, K.J. Oliver and J.T.M. **Wothers**po**o**n, Proc. Int, Conf. on Advanced IR Detectors and Systems, IEE Conference publication No 2O4 7 (1981)

(5) A. Blackburn, M.V. Blackman, D.E. Charlton, W.A.E. Dunn, M.D. Jenner, K.J. Oliver and J.T.M. Wotherspoon, IR Physics $\underline{22}$ 57 (1982)

(6) J.M. Lloyd, "Thermal Imaging Systems", Plenum (New York and London) (1975)

(7) D.E. Charlton, Mullard Ltd, private communication

(8) M.Y. Pines and O.M. Stafsudd, IR Physics $\underline{19}$ 559 (1979)

(9) T. Ashley, C.T. Elliott and A.M. White, RSRE Newsletter and Research Review (1982)

(10) T. Ashley and C.T. Elliott - Submitted to IR Physics

(11) G. Duggan and D.E. Lacklison, J.Appl Phys. $\underline{53}$(4) 3088 (1982)

(12) J.T.M. Wotherspoon, Mullard Ltd, private communication

(13) D. Day and T. Shepherd, Solid State Electronics $\underline{8}$ 707 (1982)

(14) D. Day and T. Shepherd, Solid State Electronics $\underline{8}$ 713 (1982)

(15) S.P. Braim and A.P. Davis, RSRE private communication

(16) A. Sleigh, Proc. Int. Conf. on Advanced IR Detectors and Systems IEE Conference publication No 204 19 (1981)

(17) C.T. Elliott, UK Pat 16279/78 (1978)

(18) D.B. Webb and S.P. Braim, Proc, Int. Conf. on Advanced IR Detectors and Systems, IEE Conference publication No 204 13 (1981)

(19) H.M. Runciman and S. MacGregor, Ibid p.24

(20) G.M. Cuthbertson and A.D. MacGregor, Ibid p.30

(21) W.T. Moore and B.W. Barringer, Ibid p.119

(22) I. Bleicher, N Diepeveen, D.G. Taylor and R.C. Parry, Ibid p.112

(23) J.D. Beck, M.A. Chapman, M.A. Kinch, S.R. Borello and C.G. Roberts, Proc. Int. Elec. Device Meeting, Washington D.C., p157 (1980)

(24)   R.A. Ballingall, Proc. Int. Conf. on Advanced IR Detectors and Systems, IEE Conference publication No 204   70   (1981)

(25)   I. Baker, S. Wilcock and R.E.J. King, Ibid p.76

(26)   R.A. Ballingall, I.D. Blenkinsop, C.T. Elliott, I.M. Baker and D. Jenner, Electronic Letters 18 No 7   285   (1982)